W9-BYH-047

THE LAST BIG CATS

An Untamed Spirit

Text by Erwin A. Bauer
Photographs by Erwin and Peggy Bauer

Voyageur Press

Edited by Michael Dregni
Designed by JoDee Mittlestadt
Printed in China

First Hardcover Edition
03   04   05   06   07   5   4   3   2   1

First Softcover Edition
05   06   07   08   09   5   4   3   2   1

Library of Congress Cataloging-in-Publication Data available

ISBN 0-89658-593-X (hardcover)
ISBN 0-89658-742-8 (paperback)

Distributed in Canada by Raincoast Books,
9050 Shaughnessy Street, Vancouver, B.C. V6P 6E5

Published by Voyageur Press, Inc.
123 North Second Street, P.O. Box 338,
Stillwater, MN 55082 U.S.A.
651-430-2210, fax 651-430-2211
books@voyageurpress.com
www.voyageurpress.com

Educators, fundraisers, premium and gift buyers, publicists,
and marketing managers: Looking for creative products and
new sales ideas? Voyageur Press books are available at special
discounts when purchased in quantities, and special editions
can be created to your specifications. For details contact the
marketing department at 800-888-9653.

On page 1: *The secretive cougar is more likely to be heard caterwauling during the breeding season than to be seen anywhere it survives.*

On page 2: *In the clear morning light, a cougar looks over its dry domain in the southwestern desert of the United States.*

On page 3: *An African leopard uses a stout tree in Kenya's Nakuru National Park as both a day bed and an ideal place to watch for game below.*

On pages 4 and 5: *The cheetah's long legs allow it to quickly reach top speed in a sudden rush to capture prey.*

On page 6: *The jaguar probably survives in greater numbers in Brazil than anywhere else in its vast original range.*

On page 7: *An Indian tiger explodes from dense cover to try to snare a sambar deer wading in a shallow pool.*

Opposite the title page: *The "king of beasts" gazes out over its domain.*

Inset on the title page: *A Bengal tiger growls at intruders.*

Opposite page: *A leopard mother keeps guard over her cub in the Masai Mara Game Reserve of Kenya.*

CONTENTS

THE LAST BIG CATS

An Untamed Spirit

During a lifetime spent prowling the world's wild places with a camera, one of my most vivid memories is of the incident that occurred on March 2, 1956, in what was then Tanganyika in British East Africa. That day will haunt me as long as I can still carry a tripod and a few rolls of film into my own backyard.

I had come to Africa only days before with my photo gear, limited funds, and great expectations. I had traveled across Africa before—mostly on foot as a soldier during World War II—but this was my first trip to what I regarded as the *real* Africa of adventure, wildlife, and wilderness that I had read about since boyhood.

After renting a Land-Rover that had seen far better days, I met Keith Cormack, who was a British missionary's son with a wanderlust to match my own. He had only slightly more experience in the bush than I, yet he boasted three invaluable skills. He could speak Swahili, the native tongue; he was a capable game spotter; and, maybe most importantly, he was a passable auto mechanic and knew how to fix the multitude of flat tires we eventually suffered.

Our plan was to head directly toward the famous Serengeti Plain, camping and sleeping on the ground beneath the vehicle. We had neither tent nor mattresses. We drove to the Ngorongoro crater, one of the world's largest and most spectacular calderas. Reaching the top of Ngorongoro's rim at an altitude of about 7,500 feet (2,250 m) late in the afternoon, we descended the steep, narrow Lerai Track some 2,500 feet (750 m) to the crater floor by nightfall. Although exhausted, I slept little, trying to identify the night sounds close by in the dark.

Shortly after daybreak, Keith and I drove our Land-Rover around a brooding swamp on the floor of the ancient volcano. Most mornings in this part of the world are sunny, but now a gray sky threatened rain. The gloomy scene before us could have been on another, strange planet.

Then, about 100 yards (90 m) away, we spotted a hippopotamus beside a shallow pool. Through our binoculars we saw that it was a female hiding a pink, newborn calf. Without thought or hesitation, I grabbed a camera and plunged headlong into wet, waist-high marsh grass. Not having a long telephoto lens handy meant I had to stalk up close to the hippo to get decent pictures. It was a tough, slogging approach, but I was as excited as I was foolhardy.

Suddenly, some sixth sense warned me to stop. It was not a second too soon. A lioness crouched in my path, watching me with cold, yellow eyes. Except for her tail, which flicked once, she did not move.

Winter is traveling time for cougars as they search for big game wintering areas.

A rush of terror engulfed me. Then, the faces of several other lions appeared out of the dense vegetation, surrounding me, staring at me. This pride of lions had been stalking the baby hippo when I interrupted them.

My first impulse was to turn and run, which would have been a fatal mistake. Somehow, liquid in the knees and never taking my eyes off the nearest lioness, I backed away very, very slowly. Those cold, yellow eyes watched me all the way.

Once safely in our Land-Rover, I sat for a long time. I was weak and so wet with sweat that I could smell my own fear.

Despite almost becoming a statistic, my fascination, admiration, and interest in the world's big cats really began that morning. These were the first wild cats I had ever seen outside of the city zoo in my hometown of Cincinnati, Ohio. Before that African trip ended, I watched and photographed many more lions, as well as several cheetahs, and their prey. During the nearly half century since that trip, my wife Peggy and I have visited most of the places where these ultimate predators survive.

The group of cats known as "big cats" includes some of the most beautiful creatures on earth. Humans have long been fascinated by their strength, their grace, and perhaps especially by their secretive lifestyles. We still do not know enough about them.

Most scientists regard a "big cat" as one that can roar, but not purr, because of a bone structure in the larynx called the hyoid. This group includes lions, leopards, tigers, and jaguars. Cheetahs, snow leopards, and mountain lions have a different bone structure and cannot roar. Although not technically big cats, the last three are usually included in that category by most people.

All cats share a common ancestry, belonging to the scientific order Carnivora that originated about 65 million years ago when dinosaurs died out and smaller mammals began to proliferate. Yet despite all of our modern research, the evolution of cats is not completely clear and the meaning of the fossil record is still debated. From the preserved remains excavated in California's Rancho La Brea Tar Pits, we know that saber-toothed "tigers"—Smilodons—lived in North America until about 20 million years ago, as well as in Europe and northern Africa. Cheetahs and lions once hunted rhinos and camels on the North American Great Plains; those vanished lions were a good bit larger than the lions I met in Ngorongoro. In any case, all of the large and small, household and wild cats now living belong to the same family, Felidae, which contains thirty-eight members.

Too many of these cat species are threatened or endangered. Big cats once lived almost everywhere on earth, from South America's Cape Horn northward nearly to the Arctic Circle. They are still widespread, but most are severely re-

A hungry tiger scans the water's edge to study the prey that has gathered to drink.

stricted to scattered small pockets of suitable habitat. And their numbers grow smaller all the time. Lions survive in only about one-tenth of their range of just two centuries ago. Tigers roam in less than one-fifth of their former range.

Our understanding of big cats has been hampered by the cats' very nature. Only three African big cat species—lions, leopards, and cheetahs—are easy to see in the wild, though only in national parks and other sanctuaries where they have not been hunted for a long time and where they are less fearful of humans. In addition, all big cats, except cheetahs, are primarily nocturnal hunters, which makes them much more difficult to study than other wildlife.

Our knowledge of feline lives and natural history began to improve in the 1970s with radio telemetry: By attaching radio transmitters to collars around the necks of big cats, biologists could now track the animals whether they could see them or not. Telemetry does have drawbacks, however. For biologists to put the radio collars in place, they must trap and tranquilize the animals. Other effective study methods are also now in use, such as camera traps that are triggered when a cat walks through an infrared light beam across game trails.

The big cat species share many physical characteristics. However, they vary greatly in size, from a mountain leopard of Yemen or Oman at less than 100 pounds (45 kg) to a tiger at 750 to 800 pounds (330–360 kg) and from 5 feet (1.5 m) in length for a snow leopard to 11 feet (3.3 m) for an adult Siberian tiger. Big-cat bone structures are so similar that identifying species by skeletons is not easy, and only the closest scrutiny can differentiate between a lion and tiger skeleton.

Armed with sharp teeth and claws plus keen senses, the big cats are stealthy, quick, and ferocious predators at the top of the food chain. All of them are splendid, silent hunters. All can travel and hunt without making much noise, even in a dry, brittle environment. They walk literally on tiptoes. When they run, they retract their claws, and only the soft pads on the bottom of their feet touch the ground. Powerful muscles drive their legs. During a chase, a cat is able to arch its backbone enough to bring its longer rear legs and feet up ahead of its front feet. This greatly increases the length of its stride, its acceleration, its running speed, and its ability to make long jumps.

In all the cats except the cheetah, the sharp claws retracted while running are extended on contact with prey. The claws, plus the cats' canine teeth, also grasp and hold down prey. Their long canine and shorter carnassial teeth set in short jaws are the animal kingdom's best-designed tools for killing prey larger than the cats and for eating the flesh. The side carnassial teeth grind the meat enough to be swallowed. A tongue covered with small rasps scrapes all the meat from the bones of the big cats' kill.

The senses of the big cats are as keen as any in the animal kingdom. Sensitive ears make it possible for the big cats to hunt at night, a distinct advantage over their prey. Eyes set

Although the snow leopard at times descends into lower elevations of remotest central Asia, it is usually a creature of the highest mountain ranges.

on the sides of their heads give cats a wider field of vision than that of humans. Their sense of smell may not be as keen as that of some other wild creatures, but it's adequate to follow fresh trails. Not the least of felines' advantages in hunting are the long, sensitive whiskers that allow cats to "feel" their way through brush in dim light or even total darkness.

All of the felines are lean and extremely athletic for their size. Fat cats do not exist in the wild. Powerful muscles from jaws to tail enable some—lions and tigers especially—to outrun and then batter and bring down heavy prey. Tails that can be almost as long as the body provide the balance necessary for bounding over any kind of terrain.

The secretive habits of big cats and their fear of humans have made many of them difficult, if not nearly impossible, to photograph in the wild. Lions and cheetahs are easy enough to shoot on most photo safaris in eastern or southern Africa. Leopards are a little more shy, but in many reserves they are becoming increasingly tolerant of tourists. Most trips to national parks in India produce sightings of tigers. But the most skillful, determined wildlife photographer might spend a lifetime in the field with no more than a glimpse of the other big cats. The snow leopard may be the most elusive of all; my own closest encounter was a set of day-old tracks.

Because of these difficulties, the photographs in this book of jaguars and cougars—which we have seen in the wild—and of snow leopards are of animals in controlled situations, although in natural backgrounds and exhibiting typical behavior of animals in the wild. Some of the animals were filmed in captive breeding compounds for endangered species, although some may actually be living in the wild today.

I certainly hope so.

CHAPTER 1

LIONS

The King of Beasts

 In almost every culture, Simba, the African lion, is known as the King of Beasts. The male lion is indeed regal in its bearing, its face encircled by a thick mane. Yet the male lion is also majestic in ruling over its kingdom: Unlike all of their big-cat cousins, which lead solitary lives, the gregarious lion is a social animal that lives in extended family groups called prides.

The lion, *Panthera leo*, is also the world's most thoroughly studied big cat. Of all the big cats, African lions, *Panthera leo leo*, are the most diurnal and the least secretive, and are therefore easier to find and observe than any other carnivore. Thus, some of the most distinguished animal researchers have studied lions, but the cats also continue to intrigue nonscientists as well. The opportunity to see lions living in the wild now draws more tourists to Africa than any other attraction, including the pyramids, the Sphinx, and other antiquities of Egypt. Peggy and I have made many trips to the continent to track and photograph them.

Before daybreak one morning during a safari to Kenya's Mara Plain, we drove from our tent camp beside the Talek River to follow a mostly dry watercourse toward the sunrise. But the landscape, which the day before teemed with zebras and other game, was now almost empty in the first, yellow light. We turned onto a low knoll overlooking a vast area and parked there to watch and listen. Nothing stirred. The only noise was the weak sound of a flappet lark. Then suddenly a lioness materialized from out of nowhere, looked straight into our startled eyes, and trotted off toward the horizon. Because the flat terrain made tracking easy, we decided to follow slowly at a distance.

After a half mile or so, the lioness disappeared into a jumble of rocks that formed a low ridge. We pulled up to within telephoto-lens range of the rocks and parked again. In a few minutes, she reappeared from a dark crevice, this time followed by a second lioness and then seven cubs. As three of the cubs were smaller than the others, it was evident that here were two separate litters.

While the two lionesses stretched out in the warming sunshine, the cubs tumbled, fought, and played all around and over the top of them. Soon more lions arrived at the rock pile. Two lionesses followed by a young, still-maneless male might have been returning from a successful hunt, judging from the blood covering all of their faces. A little later we heard a chorus of roars, then two large, orange-maned lions, also with reddened muzzles, joined the pride. The lions greeted one another, licking faces, chasing tails. It was the most revealing, intimate look at the family life of lions we were ever fortunate enough to see.

* * *

The low golden sunlight rays betray a lion waiting in a patch of shade for prey to graze nearby.

To tell the truth, Queen—rather than King—of Beasts might be a more appropriate title for the lion. Being exclusively meat-eaters, lions must hunt, successfully and relentlessly, to survive during drought and downpours and whether prey is abundant or scarce. Female lions do the vast majority of the hunting and killing, as well as raising the cubs, that helps the prides survive.

Late one evening while watching a lioness in Uganda stalk a zebra that had become separated from its herd, I realized how well female and male lions have evolved and developed to fulfill their different roles. Lionesses are lighter in their tawny coloring, sleeker and slimmer, more agile, much faster afoot, and more adept at concealment in even the sparsest cover when hunting prey. Males are heavier with darker manes and are less easily camouflaged. They are also more compact, all traits that are better suited for fighting and defense. The presence of both hunters and defenders is essential to the well-being of the pride.

The lioness I watched now was a study in patience and cunning. With belly scraping the ground and taking every advantage of the concealing short grass clumps between her and the zebra, she crept ahead only when the zebra was feeding, head down. Each time the zebra raised its head, she froze and became almost invisible.

Although the stalking lioness I watched was still too far from her prey to present a real threat, she did have the wind in her favor. A mistake hunting lions sometimes make is to disregard the air currents that carry their scent. But this lioness was closing in to what might be called rushing distance, and I felt my pulse pound as the drama unfolded. I could see her muscles quivering as the zebra, still unalarmed, actually took several steps closer to her. When the striped head was lowered once again, the cat rose to a half crouch, hesitated a split second to gather herself, and sprang. She was a blur of gold in the dry grass.

For an instant I thought she would score—and her long, agate claws may even have raked the zebra's shoulder. But she seemed to stumble and lose speed as the quarry, now panicked and running, sharply changed direction, its hooves making a cloud of dust into which both animals disappeared. Although swift in the short distance, lions are not long-distance runners, and this lioness quickly gave up, husbanding her physical resources for another hunt. She stood for a few moments, stared into the distance, then sat down to rest.

Not all hunting by lions involves such stalking. Especially during dry seasons in arid regions where waterholes are few and scattered, lions often linger around the water's edges in ambush. Or they skulk along the game trails that lead into the waterholes, waiting for the unwary. All animals desperately need to drink sometime, and they know that danger is lurking near their vital source of life. So they approach waterholes slowly, warily, especially where there is enough brushy cover to hide the killers. When an antelope, kudu, zebra, or giraffe cannot stand its thirst any longer, a waterhole can suddenly become either a scene of violent death, or a resting spot for another disappointed lion, panting under a hot sun.

One day I watched a herd of zebras walk haltingly to the water's edge not far from where eight lions had crouched silently, motionless since dawn. A zebra began to drink, followed by another—and then the scene simply exploded. Braying zebras bounded in all directions, spray flew, but in the end eight drenched lions walked slowly back to shore where they were still hungry when darkness fell. Fortunately the pride did make a kill the next morning as well as another soon after that.

Many prey species are faster afoot than lions, and others are too formidable for a lion to tackle alone. This makes cunning and cooperation as well as brute strength important in the hunt. Lions work together in hunting just as they do in the communal caring for their young, separating them from all other felines, which avoid their own kind except to breed. Two or more lionesses may circle out of sight to positions downwind of a vulnerable antelope. One or two of the fastest, most-experienced hunters in the pride patiently and stealthily stalk the target, using any cover, defilade, or shadow to their advantage. From a point as close as possible to the prey, the lionesses make the final rush. Just often enough, the startled target turns and races away directly toward the other lionesses waiting downwind. So while lions may individually be less skillful as hunters than a single leopard that shares their range, the end result is the same: The kill is made.

A healthy adult lioness can bring down some prey with a single swipe of her paw and crunch it to death in her powerful jaws. When one lioness does manage to catch a large animal by cunning, stealth, and surprise, she desperately hangs on with claws and teeth until it falls over or help arrives. Whenever possible, she tries to grab the throat or nose and then to hold on until the victim suffocates.

Few lions are able to hunt and score consistently until they are about two years old. Even the most experienced cats average several attempts for each successful kill during daylight hours, usually in early morning or late-afternoon forays.

During a thirteen-year period in the 1980s and 1990s, biologist Dereck Joubert and photographer Beverly Joubert, a South African husband-and-wife team, studied the lions of Botswana's 4,000-square-mile (10,400-sq-km) Chobe National Park. Since other scientists had done research during daytime, the Jouberts decided to work at night when they knew lions were most active. They studied about 120 animals in six separate prides in the Savuti area, logging more than 24,000 hours of observation. They followed about 4,750 hunts for quarry, from impalas and wildebeests to Cape buffaloes and even an elephant. The lions were successful in fewer than 30 percent (1,300) of these hunts. They made only 5 percent of all the kills in daylight. They preferred to hunt on open grasslands, and had the greatest success when the moon was half full or less. The hunters were rarely successful or even active beneath the light of a full moon.

For this hunting party of Kenya lionesses, the chase ended when their warthog escaped into a deep underground den.

It was evident to the Jouberts, as well as to other researchers, that the lion's strategy of group hunting makes all the difference between success and failure. Of course, single lions sometimes do make kills, especially of young or old prey. But the odds of success multiply dramatically when a hunter is able to flush a target toward members of the pride waiting in ambush. Although the meal must then be shared among a large number of hungry pride members, the large size of the adult prey satisfies many appetites for a longer time. Lions may not actually plan such strategy, but their instinctive cooperation compensates for the killers being slower afoot than the victims. It also permits a pride of lions to attack and kill prey, even adult buffaloes and rhinos that individual lions could not subdue alone.

A lion pride may consist of as few as five or six animals to as many as twenty-six, as seen in one pride studied in Tanzania. An enormous pride of thirty-five lions was once counted elsewhere in that country.

The adult lionesses are the core of every lion pride. Nearly all of them are born, live, mate, and die without leaving their sisters, mothers, aunts, and cousins in the same pride. Some lionesses may leave with their kin to split off into new prides, or to live nomadic lives, but they do not join other existing prides. All male lions, on the other hand, eventually separate from the prides of their birth to establish families of their own, thus avoiding inbreeding.

All of the females in a pride come into estrus at approximately the same time; many biologists believe that this is when the prey supply is most abundant. The dominant male repeatedly mates with every lioness in a pride; this is an important role of the male, and in a large pride it can be exhausting. At times it even seems a chore to the male, and is thus maybe initiated by the female. Mating takes place for the two to four days when the lionesses are receptive. A tryst begins when a lioness comes into estrus and walks in tight circles around the pride male, flicking her tail near his face to give him her scent. She then entices him to follow her a short distance from the pride. The breeding pair may spend two or more days together, grimacing, snarling, biting, lolling side by side, but mostly simply resting between couplings, which are frequent. During research in Etosha National Park, Namibia, biologists Des and Jen Bartlett observed one pair mating twenty-three times in less than six hours. The male was the aggressor only once.

Approximately three and a half months after mating, and at roughly the same time, each female in the pride bears one to five cubs. The cubs are weaned within six months.

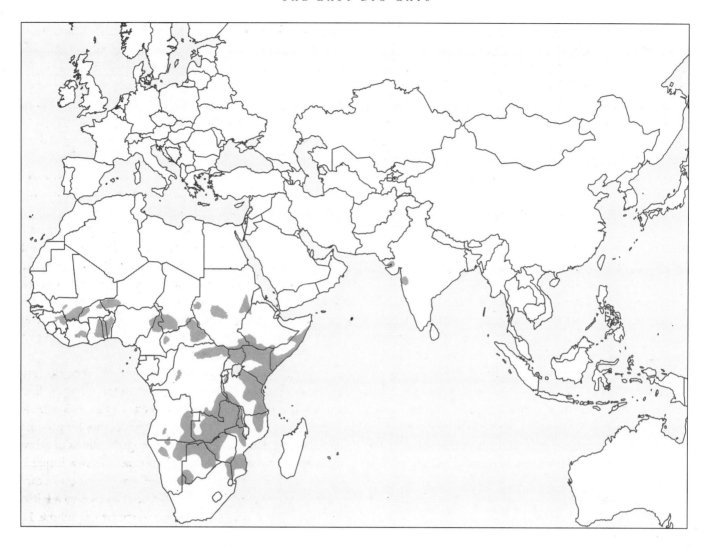

Above: *Range of the lion*, Panthera leo

Left: *Track of the lion*, Panthera leo
Front adult paw length: approximately 4½ inches (11.25 cm)

The greater the supply of game, the easier it is to nourish all of the pride's cubs, and communal rearing is possible within a small, guarded territory.

If one lioness does bear a litter much later than other pride females, she usually tries to raise them alone, away from the pride. She is seldom successful, however. Too often, hyenas, wild dogs, jackals, leopards, or cheetahs find the young while the single lioness is away hunting. Cape buffaloes have been observed trampling lion cubs to death, perhaps deliberately. Still, these late cubs would have had a hard time competing for food with the older, larger cubs in the pride if they had remained with the main group. In either case, their future is in serious doubt.

Life is more hazardous for male cubs than for females. If a new pride male does not kill the previous male cubs early on, they are evicted when approaching adulthood. This is the critical period when many males perish. But if they are tough enough to make it alone or as one of a two- or three-bachelor group, they may eventually take over a pride of their own. When they do so, few small cubs from litters prior to their takeover survive: All cubs are either abandoned or killed by the new dominant male or males. With no cubs to nurture, the lionesses come into estrus promptly, and the new males sire new cubs.

Brutal as it may seem, this also contributes to the long-term survival of the species as the progeny of the takeover males are likely to be stronger and healthier than the cubs of the losers. A male can maintain dominance in a pride for just two or three years at most before being replaced. American biologist George Schaller—whose fieldwork has lured him from studying mountain gorillas in Africa and tigers in India to jaguars in South America, snow leopards in the Himalayas, and giant pandas in China—began his extensive field studies among the lions of Tanzania's Serengeti Plain. In one Tanzania pride, Schaller recorded nine sets of breeding males in seventeen years of observation.

At first it is easy to regard a male lion as simply a handsome yet lazy, useless parasite on the pride. Most males we have observed over the years either spent their time sleeping, or feeding on prey killed by lionesses or even hyenas. But the poor reputation is not deserved. Besides simply functioning as the father of the cubs, the male is the defender of the territory on which the entire pride depends for survival. In any region where there is suddenly a surplus of wandering males, this defense is a full-time job that can involve savage fighting. By defending a territory, the pride male is also providing a secure place for females to raise his cubs.

The male's mane plays an important role in the lion's life. While a long mane that blows easily in the wind may be difficult to hide when hunting an alert zebra, it is a distinct advantage in defense. When patrolling the border of the pride's territory and marking it with urine, the dominant male can be easily seen by any intruding males, and that sight alone might well convince an interloper to move on. That same mane gives his neck a good bit of protection in case the intruder does not turn away and chooses to challenge the resident lion for his position. No matter what the size of the dominant male's mane, territorial fights are far more likely to occur when the aura of estrus is hanging in the air.

The roar of a male lion is unlike any other sound in the world. Driving back to my camp in Namibia in the dark one night, I heard a lion roaring in the distance. Again and again the sound reverberated across the night. In all wild Africa, no sound is more stirring and evocative, but on this particular black, moonless night when I was alone, it especially stirred—and even chilled—my imagination. Exactly where was this lion, and why was it roaring?

The consensus is that lions, usually the males, roar to advertise their presence or their territorial claim to any intruders. They may also roar so that other pride members know where they are. Lions almost certainly recognize each other or identify strange lions by their distinctive roars. Both native trackers and wildlife biologists have assured me that different lions have different voices and repertoires that can easily be recognized.

A typical roar is a series of penetrating bass grunts rising in crescendo before slowing. Low-intensity roaring may be a lioness calling for her cubs to come out from a hiding place. Humans can typically hear a lion's roar from about a mile away, but other lions can hear the roar from much farther than that.

On one occasion, the loud roaring of a lion early in the morning led me to a waterhole on Kenya's Mara Plain where a small pride was just abandoning the remains of a wildebeest kill to the gathering jackals and vultures. The lions' stomachs seemed full, which was well and good because days might pass before they could gorge like that again. Lions are adapted to a feast-or-famine life, and when food is plentiful, they take full advantage of it. Schaller reported that five lions once consumed a 600-pound (270-kg) zebra, eating its meat, hide, and entrails—all but its bones and stomach contents—during just one day. He also watched one male eat 73 pounds (33 kg) of meat—one-fifth its own body weight—in a single night.

On the East African plains, 90 percent of the lion's food consists of just two species: zebras and wildebeests. Lions will eat anything they can capture, however, from warthogs to ostriches. We have watched a whole lion pride trying to dig a warthog out of its hole. Kills of adult hippos by lions have been recorded in South Africa's Kruger National Park and the Central African Republic's Virunga National Park. Although normally considered meat eaters, lions in Kruger have been observed feeding on barbels and other fish in streams and pools that were evaporating during dry seasons. Most creatures seem fully aware of the lion's capabilities and sense the distance to which a lion can approach without posing a genuine threat: The rule seems to be that a visible lion is a safe one, even if it's close by. Most animals also seem able to sense whether an approaching lion is a hunter and hungry, or simply traveling through.

Scavenging from other predators and eating carrion are common lion behaviors as well. During his Serengeti lion studies, Schaller continuously followed one past-prime male over its 75-square-mile (195-sq-km) territory for twenty-one days. During that time, the lion scavenged a chunk of zebra from several hyenas, retrieved a half gazelle from a leopard, and joined his female companions dining on an eland carcass. The only time this lion tried to hunt on its own it failed to catch a dik-dik, a tiny antelope weighing only about 10 pounds (4.5 kg).

During one rainy season, Schaller recorded lions feeding on a total of 121 different carcasses, and he inspected each kill carefully. Only half the kills had been made by the lions feeding on them; most of the rest were stolen from hyenas. During the same period he saw a pair of lionesses hurriedly feeding on their kill while hyenas assembled all around them. When fifteen howling and moaning hyenas had congregated, they finally advanced on the larger cats and drove them away. This is not uncommon in areas where spotted hyenas greatly outnumber lions.

In scattered areas of their range, lions have learned to climb trees for a variety of reasons. Some lions scale trees to scavenge cached leopard kills. In Tanzania's Manyara National Park, several prides spend midday hours above ground in thorn trees, probably because it is cooler and free of bugs there. But they have also been observed escaping into the same trees when confronted by herds of snorting buffaloes. Females in one Manyara pride seemed to climb trees for a better view of the game grazing around. By contrast, another female survived and fed cubs by catching injured flamingos from the vast flock that lives and nests on Manyara's lake.

When it comes to feeding time, it's every lion for itself. The warm scenes we have photographed of cubs gamboling with their mothers suggest that lionesses are protective, caring, and tolerant parents. But this is not always the case. Having made a kill, the lioness or lionesses responsible will eat as

Lions once ranged widely across all of Africa, southeastern Europe, and southern Asia, but this Indian lioness is one of only a few hundred survivors outside Africa in India's Gir Forest.

as the snarling females fought over the prey that lay limp and inert in the grass. Not until the carcass was completely torn apart did the lionesses walk away to fetch the cubs that were cached nearby. When they returned, they found the rest of the pride guarding what little was left, and these slashed out at any of the cubs that dared come near. The incident caused Schaller to wonder if the lioness wasn't sometimes an unfit mother.

By human standards this behavior may not paint an attractive picture of the lion, but we cannot judge the species the way we would other humans. In order to survive in wild Africa, the strongest simply have precedence. Whether one considers the lionesses unfit mothers or not, the fact is the overall system has worked for ages. Schaller made his observations of this pride in 1966; two decades later, descendants of the pride were still living in and defending the same territory.

In the not so distant past, Africans still hunted lions with primitive spears, knives, and buffalo- or giraffe-hide shields. Often, the hunters were dealing with livestock killers, but sometimes this hunting was simply a ritual or test of manhood. The tall, slim Masai of southern Kenya and northern Tanzania were the best known and perhaps the bravest of the lion hunters, and their exploits have been commemorated in museum displays, bronze sculptures, and paintings.

After a lion killed a Masai cow, the young *moran*, or warriors, of the community would assemble. Following frenzied dancing to build courage, they tracked the killer to its resting place, which was probably in dense bush. There the ring of men carrying spears, shields, and *simis* (short knives) surrounded the lion. The tormented cat would strive to break out of the tightening circle of warriors, almost always toward the least-experienced or least-determined *moran*. Its escape would be halted by a hail of spears. The cat was then normally finished off in a bloody mêlée of men and knives. Fearful wounds were inflicted all around. The warrior who was able to cut off the lion's tail while it was still alive was thereafter regarded as a superior by his peers.

Being carnivores, some lions inevitably become man-eaters. Over the years, man-eating lions have terrified Africa from Nairobi to Cape Town. These lions have become both famous and infamous.

One of the most notorious of these was a light-colored cat nicknamed Chiengi Charlie that terrorized northern Rhodesia (now Zambia) in the early 1900s. This lion would cut down an African in one place, disappear, then subdue and eat another miles away, eventually taking a terrible toll of hundreds of humans. The colonial government went to great lengths and spent a great deal of money to shoot this menace. For a long time, the man-eater easily eluded all.

The area's top professional hunting guide was engaged to kill Chiengi Charlie, but the cat dragged away the hunter's

much as quickly as they can before being driven away by more-dominant pride members, usually the males. Desperately hungry members of the pride may even begin feeding before the captured creature has drawn its last breath. The smaller cubs are always the last to feed, if any meat remains at all. This is just one reason why the mortality rate of lion cubs may exceed 75 percent. Another reason is that lionesses are not ferocious protectors of their cubs from the dangers presented by predators as are, for example, grizzly bear mothers.

In the Serengeti, Schaller studied one particular pride that he discovered one day, sprawled together in the shade of an acacia tree. It was that vision of the idyllic Africa that travelers and poets never forget. Pride males and females rested together, Schaller recalled, in a tawny mass, bodies touching. The cubs played with the pride matriarch's tail and quarreled with one another.

Several nights later in the light of a pale moon, Schaller found several lionesses of that same pride that had just pulled down a zebra. Now the air was heavy with the scent of blood valet while his employer was taking a bath. In time, two other lions joined Charlie, making a deadly threesome, and they almost depopulated one entire rural district as survivors fled for their lives. In desperation, a series of gun traps were set over a wide area; after also killing several unsuspecting farmers and their cattle, the traps finally brought down two of the killers. The man-eating ended with that.

Some areas of wild Africa seem to be man-eater prone. During the early 1930s, most of the wildlife in the Sanga area of Uganda was destroyed when the bush was burned to prevent a spread of rinderpest to livestock. Surviving game herds migrated to greener pastures, and without their natural prey, the lions turned to eating humans, which they continued until the 1970s. Along the main road from Sanga to Rwanda, one lion accounted for eighty-four victims; another killed at least forty-four. Man-eating recurred there in 1937 and 1938 when several prides became involved in wholesale killing of villagers. A year and a half of concentrated hunting by a company of game scouts was required before all of these lions were dispatched.

Another Uganda lion developed the habit of following elephant herds, apparently noticing that people left the safety of their homes to chase away the elephants invading their gardens. That made it easy for the man-eater to catch an easy meal. But this cat made a fatal mistake one morning when it tried to catch an armed game warden also tracking the elephants.

During the railroad-building era in Africa in the late 1800s, lions stalked the railway crews. They killed thirty crew members working on the tracks from Beira, Portuguese East Africa (now Mozambique), to Salisbury, Rhodesia (now Harare, Zimbabwe). Man-eaters also were a terrible problem along the Eastern Line Railway through the Transvaal of South Africa as workers laid the first tracks through bush country.

The man-eaters of the Tsavo area of Kenya are perhaps the most famous killers, a pair of maneless male lions that, in 1898, actually halted construction of the Mombasa-Uganda Railway. In an almost nightly reign of terror, the pair dragged away a total of twenty-eight Indian workers and dozens more African laborers from their barricaded quarters. Finally, all work had to be stopped.

The Tsavo lions developed an uncanny, almost magical ability to enter any dwelling to capture a victim. Working crews lived in constant terror. Finally, Lieutenant Colonel John Henry Patterson, an engineer in charge of railroad bridge construction, shot the lions. In 1927, Patterson wrote a best-selling book about his adventures, *The Man-Eaters of Tsavo and Other East-African Adventures*. The lions' skins were shipped to Chicago and may be seen today in the Field Museum of Natural History.

A male lion roars in response to another lion roaring in the distance.

Patterson didn't completely stop the man-eating along the railroad, however. Other workers kept vanishing as the rails pushed farther upward through the wilderness. Near Voi, Kenya, a lion dragged a European engineer from his tent and dined on his body in full view of workers huddled in their construction camp. Professional trappers arrived to dispatch the animal for the $500 bounty, a large sum a century ago. In 1955, lions were still harassing station workers, killing a telegraph operator near Tsavo as he wired for help.

It is often written that man-eaters are invariably old, perhaps mangy or sickly animals no longer able to catch and kill their natural prey. Sometimes that is true. Chiengi Charlie was not a young, vigorous lion. Another cat that ate seventy-five people near Mgori, Tanganyika (now Tanzania), in 1958 was finally shot by a professional hunter who examined the carcass closely. Its teeth and lower jaw had been smashed and rendered almost useless long ago by a muzzle-loading shotgun ball. The cat had therefore turned to catching porcupines until its paws and muzzle were so full of quills that there was nothing else left for it to eat—except people.

Yet just as many of the man-eaters were healthy and in the prime of life, according to naturalist C. A. W. Guggisberg, who spent many years researching the species. The Tsavo cats are an example. One of the most recent man-eaters shot in Barotseland, Zambia, was a huge male in excellent condition. In fact, its lustrous pelt, which measured almost 10 feet long (3 m), is a large one for the species.

The deadliest spree of man-killing by lions ever recorded occurred as recently as the 1990s. For more than a decade, Mozambique descended into a nightmare of civil war and terror. Many thousands of Mozambicans attempted to escape into South Africa, which meant walking across Kruger National Park, a wilderness where lions are numerous. No one knows how many hapless illegal immigrants the cats killed, but they probably numbered in the hundreds. One pride of five lions had to be euthanized by park rangers because it stopped hunting animals and specialized in the far-easier human prey.

In 1998, tourists in one section of Etosha National Park, Namibia, often encountered an emaciated, fourteen-year-old lioness and her companion, a scruffy, bag-of-bones male. Somehow the two managed to survive at a waterhole where the male flushed game toward the female. When the waterhole and supply of game dried up, the pair staggered slowly over parched land to another waterhole 25 miles (40 km) away next to Okakuejo, a popular tourist camp. The same night the starving pair arrived, a tourist decided to sleep outside his tent and under the Southern Cross to experience the real Africa. It was his last mistake. Somehow the weakened lions had managed to scale a 10-foot (3-m) barrier around the camp and eat the camper. It was a case of man-eating in desperation, for which the lions were destroyed.

Now and then man-eaters still stalk wild Africa, but they are dealt with long before they can claim a large toll.

Most of the time the lion is content to live as lions have always lived in some of the most beautiful wilderness left on earth, avoiding people. Wherever there are large herds of ungulates, the African lion serves as a valuable balance of nature. From time to time illegal hunting—poaching—has reduced lion numbers, but nothing compares to the degradation and loss of habitat that expands day by day.

One thing is absolutely certain: Wild Africa would be a drearier, less-exciting land without Simba and his powerful roar. East Africans translate this roar into "*Hii nchi ya nani? Yangu! Uyangu!*" Swahili for "Whose land is this? Mine! Mine!" But of course it isn't, no matter how powerful or how threatening the lion may sound to those sleeping out on the savanna at night.

Lions once lived over much of the world. A lionlike, saber-toothed creature roamed North America during the Pleistocene epoch. Remains of this animal, estimated to be about 14,500 years old, were found in the Rancho La Brea Tar Pits of southern California and in the Friesenhahn Cave of Texas. As recently as two thousand years ago, lions in their present form lived in southern Europe from southeastern Spain, Italy, and Greece to Asia Minor, most of the Middle East (except Arabia) and Iran eastward to Pakistan, and most of India. They also occupied many kinds of habitat over the entire African continent. Hieroglyphic records in the British Museum reveal that Amenhotep III of Egypt killed more than 100 lions between 1411 and 1375 B.C. According to the Bible, Samson was a lion hunter before his famous shave and haircut. Homer of *The Iliad* and *The Odyssey* wrote of lion hunting as sport in Greece. The Colosseum of ancient Rome was the scene of battles between lions and gladiators as well as of executions of Christians, who were fed to captive lions for entertainment. Later, Saint Louis of the Seventh Crusade amused himself by hunting lions on horseback when he ran out of infidels to skewer. And the hunting of lions from horseback was a common sport of the Touareg in the Niger region of Saharan Africa.

Trophy hunting for male lions is still permitted in a few African nations, perhaps unwisely. One of these is Botswana where, in the Okavango Delta, permits are issued to shoot twelve large males annually. Under pressure from professional hunting services, that quota was raised to thirty per year in 2000.

Botswana Wildlife Department biologist Pieter W. Kat believes that number is far too many. With no more than eighty or ninety adult males in the open-hunting territory, Kat believes that many males will be attracted from reserves into the hunting land. This disruption, pride takeovers, and natural mortality may be more than the population can stand

Male lions, probably siblings, share leadership of a Kenyan pride by frequently bonding.

and will undermine the future of Okavango lions. In fact, Kat believes, lion numbers are quietly plummeting all over Africa, and lions should be regarded as an endangered species.

Almost all of the lions left in Africa are now concentrated on the savannas and lightly timbered lands in national parks and reserves where, for the time being, their future seems secure. In 1999, biologists were surprised to find a small population of "jungle lions" surviving in a remote section of the Congo Basin, one of the few still unexplored areas on the continent.

Outside Africa, lions still live wild in only one, small enclave. An estimated 100 to 200 Asian or Indian lions, *Panthera leo persica*, inhabit about 400 square miles (1,040 sq km) of semi-arid scrub in the Gir Forest of Gujarat state, western India. They share this sanctuary with thousands of cattle and are in constant conflict with the livestock owners who live in lion-proof shelters. If not for their great value as tourist attractions, these last of all Asian lions would survive only in zoos.

Lions, African as well as these Asian few, are not the victims of hunting, as is so often claimed. The greatest threat to the lion's survival is increasing human overpopulation and expansion, limiting the wild lands lions need. If our current growth continues apace, lions may not long survive. The King—and Queen—of Beasts are steadily, slowly, losing ground, like too much other wildlife around the world.

When prey is plentiful enough, lions spend the greatest part of every day sleeping, resting up for the night's hunting still ahead.

The manes of male lions range from tan and tawny brown to black, including red manes like this Kenyan alpha male.

Above: *A vulnerable herd of zebras drink at a shallow water hole in Etosha Pan, Namibia.*

Right: *Cape buffaloes are never more exposed to hunting lions than when traveling to and from familiar water holes.*

Opposite page: *A lean lioness with muzzle already bloodied in an unsuccessful hunt stalks wildebeest drinking at a water hole.*

Both photos: *A pair of lionesses stalks prey in Kenya in the dawn's golden light.*

Top left: *Lions may follow wildebeest herds during migrations but the antelope are most vulnerable around isolated ponds.*

Center left: *The oxpeckers clustered on this warthog's back serve as a delousing corps, skillfully removing parasites while ready to warn of lions lurking nearby.*

Left: *Impalas are important prey at and away from water sources for all of the African big cats.*

Above: *Despite being constantly alert, the impala is a potential meal for lions, leopards, cheetahs, hyenas, and hunting dogs, especially in the low light of dawn and dusk.*

A lioness with a litter of cubs to feed has just captured an impala fawn. Such young antelope are especially easy to catch.

A hunt for a zebra at dawn has been successful, and now most of the prey is already eaten. The rest of the day can be spent sleeping.

Both photos: *Lion family life is not always serene. The pride females have made a kill, and here the pride males come to claim the first—and best—portions.*

A zebra can be quickly consumed by a large pride of lions. Little is wasted. Here, a young male gnaws the tough meat from a zebra haunch.

A large male lion succumbs to midday heat and a full stomach to sleep the day away.

Both photos: *At daybreak the young male lions killed a Cape buffalo at a water hole. They had eaten almost as much as they wanted when a pack of spotted hyenas drove them away and scavenged the rest. The lions retreated to shade nearby to groom one another and then fall asleep.*

Top: *A baby lion interrupts the afternoon siesta of a tolerant aunt or mother.*

Above: *Almost from birth, baby lions have a curiosity for everything around them, a valuable trait for survival in their savage world.*

Storm clouds over the Serengeti Plain signal the onset of the rainy season and the great migration of plains wildlife.

LEOPARDS

The Supreme Predator

It is possible to look straight at a leopard in the wild and never see the big cat as long as it remains motionless. Once, waiting with our cameras beside a waterhole in Sri Lanka, Peggy and I sat within a few feet of a leopard. The cat was crouched in the bush and watching us all the while with its wary eyes, but we did not see it until it nervously flicked out a red tongue. Then it stood and calmly stalked away. Another time, in Kenya, a leopard in a treetop betrayed its position to us only when its tail carelessly dangled and swung in the wind. At first, the extraordinary beauty of the leopard's famous coat might seem to be a handicap, but actually it is an asset: When the leopard is lurking in dry grass or resting in the crown of a thorn tree, its spot pattern provides perfect camouflage.

Leopard skins are as highly prized by African tribal leaders and shamans as they once were by movie queens in Europe and America. Fortunately for the species, the spotted-fur coat fad seems to be history. Yet in Africa, the leopard's pelt is part of its mythology. Probably because they find the leopard's tracks far more often than they ever glimpse the track-makers themselves, many African bush people credit leopards with the ability to reason.

My first encounter with a leopard may have borne this out. In 1960, I arrived in Colombo, the coastal capital of Ceylon (now Sri Lanka), during the February dry season, and with a driver-guide, set out for Wilpattu National Park, exactly halfway around the world from my home in Ohio. I was there to try to photograph the beautiful spotted leopards for a magazine assignment. In Wilpattu, I headed for an area of 377 square miles (980 sq km) of forest classified as dry jungle. The tract is punctuated with twenty-seven *wilus*—ponds—from 5 to 80 acres (2–32 hectares) in size, alongside of which stand rest houses and small bungalows for visitors. I soon learned that leopards had established territories that included one or more bungalows or camps. I also learned that the cook at our Kali Vilu camp had attracted a herd of wild boars to the area by regularly feeding them table scraps. And of course the swine regularly attracted one of the big cats, known as the Kali Vilu leopard, in search of a meal.

I was camped in one of these Ceylonese jungle bungalows for the night when a sound outside awakened me. For several moments I sat uncertainly on the edge of my cot, wondering what the noise was. A scarlet sun was rising beyond the paneless, screenless window. Insects buzzed around my mosquito net. Then I heard the strange, grunting growl outside again—and it was coming closer. The growl sounded like the rasp of a dull crosscut saw against dry timber. Then my grogginess cleared.

A stout, tall tree in a green acacia forest is an ideal place for a leopard to rest, scan his domain, and watch the game trails passing below.

That growling sound outside was the Kali Vilu leopard. Hurriedly, I pulled on pants, tried to force a left shoe onto my right foot, gave up, grabbed a camera instead, and rushed outside. But I was too late to use it. The crafty Kali Vilu cat had already disappeared into the tall grass nearby, although I could still hear its growling sound moving away.

The English writer Rudyard Kipling, who spent much of his life in India and Africa, seemed to understand the personality of the leopard best. Almost a century ago, he described an encounter with a leopard in British East Africa: "The fringed lips drew back and up, the red tongue curled; the lower jaw dropped and dropped till you could see halfway down the gullet, and the gigantic dog teeth stood clear, terrible as the Demon." Kipling also wrote a story for children about the big cats entitled, "How the Leopard Got its Spots," in which the animal is portrayed as shrewd and cunning and wise.

Leopards are remarkable killers. Small in size compared to lions, leopards often hunt prey two or three times their own size. Disguised by that beautiful spotted fur coat, they are fiercely efficient predators, equipped and camouflaged to survive in forests, mountains, and deserts, as well as in the bush, where they compete with other accomplished killers, including cheetahs, hyenas, lions, tigers, and wild dogs. No wonder so many hunters have long considered leopards among the most desirable of big-game trophies.

Full-grown male African leopards, *Panthera pardus*, average 130 to 140 pounds (58.5–63 kg), with an exceptional one reaching 180 to 190 pounds (81–85.5 kg). They stand about 30 inches (75 cm) at the shoulder and average about 6 feet (180 cm) from nose to the tip of their tail. Adult females are smaller, weighing from 70 to 125 pounds (31.5–56.25 kg), or roughly a third of what a lioness sharing territory would weigh.

The leopard was and is still today the most widespread of all the big cats, being native to a great variety of habitats over most of Africa and Asia. Leopards live in all but the emptiest deserts as well as dense, saturated jungles; in habitats from open savannas and acacia woodlands upward to the snowlines in mountains from Kilimanjaro in Africa to giant panda country in China. Peggy and I have seen leopards in eight countries, although much too briefly in half of these. Maybe the most surprising sighting of all was of a spotted cat strolling a sandy ocean beach in Ruhunu National Park in Sri Lanka. Another was of a mating pair high above a roadside in northeastern Iran during a howling windstorm.

Wherever they have been hunted, either as trophies, for their exquisite pelts, or just as livestock killers, leopards have evolved into the most cunning and elusive of carnivores— and, in fact, of all larger animals. At the same time, they have adapted to survive, to live with humans, even among humans, as can no other animal their size. There is evidence of leopards living, even thriving, on the fringes and in slums of cities from Nairobi, Addis Ababa, and Kampala to Assam and Rangoon, where they live on rats, house cats, stray dogs, even human derelicts, or anything abroad at night.

Despite great efforts to trap and poison them, some leopards have become adaptable and live anonymously in unlikely places. Opportunistic individuals even turn up in places stranger than city suburbs, as for example in a penguin rookery on South Africa's Tsitsikamma Coastal National Park. There, in 1986, a single cat materialized to eat the penguins almost as fast as they could waddle ashore to try to establish a breeding colony.

In recent years as tourism has increased in African game parks, leopards have become increasingly accustomed to the travelers who find them sleeping, but somehow still watchful, during daylight hours in tall grass, amid rock *kopjes*, or stretched out on tree limbs in dappled-leaf shadows.

But when night falls or prey ventures near, a leopard is suddenly alert and transformed into the best athlete on Earth. It can glide from cover to cover as secretly and as smoothly as a snake. It is so perfectly coordinated, so swift afoot, that a running leopard is only a blur to the human eye or camera. Only the cheetah may be faster in a short dash, but that's because cheetahs chase prey in the open and often during the daytime.

Leopards are good enough swimmers to cross any river in their path. They are powerful enough to cache an antelope carcass so high in the fork of a tree that it is out of reach of other meat-eaters. Once in the Serengeti, I watched as a small female leopard tried unsuccessfully to drag a topi, which likely weighed 150 pounds (68 kg), into a high fork of an acacia tree. When the leopard's grip failed and the antelope fell back to the ground, the leopard and her half-grown cub stood guard nearby all day long, driving away vultures and a jackal. Finally, the pair had to watch sullenly from a limb just overhead when a pride of young lionesses appeared and appropriated the prize.

Leopards rely on stealth and surprise to catch a meal. In addition, a leopard can leap 30 feet (9 m) without apparent effort—a considerable advantage in capturing the baboons, monkeys, hyraxes, bush pigs, ground birds, and smaller antelopes that seasonally make up the bulk of their diet.

Leopards also have natural radar. The bristle tufts on their forelegs and their long chin whiskers are tactile sensors that flash information to the brain. The spotted cat's sense of smell is keen enough, but its hearing is as acute as the most sensitive electronic sensor, as many human hunters waiting in blinds have learned. Researchers and many Africans agree that a leopard's vision is second only to that of the high-flying vultures that make a living by spotting leopard and lion kills from miles away.

All cats walk on cushioned pads and attack with sharply pointed claws. A leopard has five claws on each forefoot and four on each hind foot, all equipped with retracting mechanisms. This keeps the claws from being blunted while walking, as are those of bears and dogs, whose claws do not retract. In a normal walking attitude, the leopard's claws are

The female leopard known as Halftail shares a meal with her cub. This may be the beginning of the cub's weaning.

off the ground and protected by a sturdy sheath. When the legs are outstretched to climb or strike prey, the tendons automatically unsheath the claws.

If you spend enough time on safari in Africa, you will eventually find mating pairs of lions, maybe many of them if the season is right. But it wasn't until the last day of a photographic trip to South Africa's Kruger National Park that I had my first sighting of mating leopards. It wasn't exactly what I had expected. A study in bored indifference, the male seemed to sleep while the smaller female cavorted playfully about, rolling, flexing her body, and making other advances. Eventually the male was aroused enough to rise, mount briefly, and walk away. Since then I have learned that this may not always be typical leopard mating behavior.

Other researchers and observers have offered greatly differing reports on the mating behavior of leopards. Most agree that it is a tense, even violent and bloody affair. During two years of filming the cats in the Mala Mala Game Reserve in South Africa, where the numerous leopards are habituated to the daily sounds and smells of vehicles plowing through the brush, cameramen Kim Wolhuter and Dale Hancock recorded pairings that resemble animal soap operas more than simple reproduction. Females might suddenly desert mates to seek out and follow other males roaring in the distance. Other females might leave their immature cubs on their own for trysts that last three days to a week. And still others may mate with more than one male during an estrus period while they still have young, dependent cubs.

Gestation has been calculated to be from 90 to 110 days, after which one to four, but usually two or three, cubs are born. They are weaned in three to four months and are independent a year later. Leopards are among the wild mothers that catch and carry live animals, often baby antelopes, for the cubs to "play" with. The torment is not a pretty sight, but in the process cubs begin to develop the hunting and killing skills they will need to survive.

Occasionally, the typical spotted female leopard will give birth to an all-black cub along with the two or three spotted siblings in her litter. When fully grown, these singularly glossy, melanistic animals, often called panthers, seem even more striking, more enigmatic than their kin. Melanism is a phenomenon in which dark pigment invades the hair, and it occurs in many wild species, especially cats. In leopards, the normally light hairs seem to be dyed dark brown or black, and if you look closely at a melanistic leopard, it is possible to see the darker spots. Melanism occurs far more often where leopards live in dense tropical forests and is most often found in Asia, especially in Malaysia. The sight of a black leopard anywhere in Africa is a rare event.

Whether black or spotted, leopard cubs are able to hunt and survive on their own much sooner and much more efficiently than lion or cheetah cubs. This is another reason the species is faring better in Africa now than most other

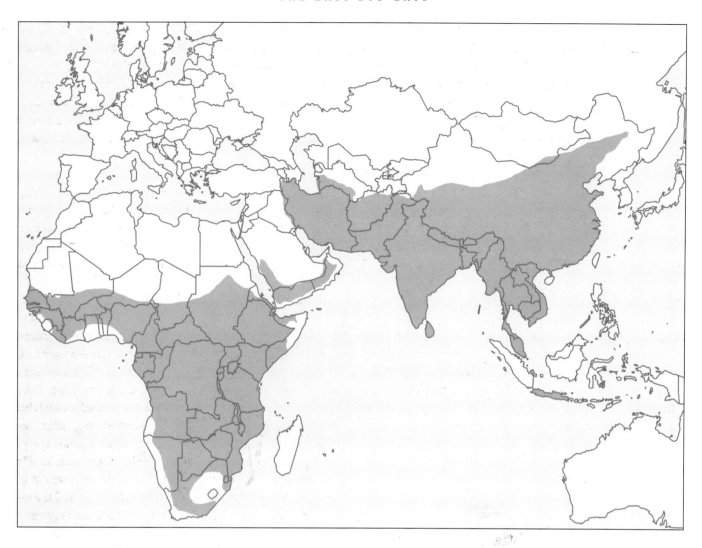

Above: *Range of the leopard,* Panthera pardus

Left: *Track of the leopard,* Panthera pardus
Front adult paw length: approximately 3⅝ inches (9.625 cm)

carnivores. Rarely do cubs remain with their mother longer than six months. As an old African tracker assured me beside a campfire years ago, by the time a leopard is a year old, the only thing this consummate hunter cannot do is fly.

The rhythm of a leopard's life is much the same wherever it lives. It clearly is among the most difficult of the big cats to study, except in a few sanctuaries around the world where all the vehicles carry tourists with cameras instead of hunters with firearms. Even in these refuges, the leopard's elegant spotted coats make it difficult to follow even in sparse cover. Somehow a leopard blends into tall sere grass almost as well as in the mottled shadows of a treetop. More than one of the finest trackers employed on African safaris confided to me that this most catlike of cats was by far the most difficult game for him to point out—or to hear. Especially when it is hunting, a leopard is a study in silence.

Leopards are predominately solitary animals. No other animal is ever anything to a male leopard—except perhaps a meal. Male leopards consort with their mates for only a few days and never see their own cubs. Leopards come together to mate, to defend a territory, or, if they are females, to tend young. As with the other cats, females live in territories that vary in size according to the abundance of prey. Male territories are larger and overlap those of a number of females. Males also seem to spend more time in the territories of the most skillful female hunters, perhaps with higher hopes of sharing a meal.

Leopards hunt at any time of day or night, apparently with the same success ratio. Overall, their kill ratio is higher than that of lions and tigers. And African leopards will hunt and eat anything from guinea fowl, sand grouse, bustards, and reptiles to hares, baboons, and antelope of all ages. Once driving at night in Kenya's Nairobi National Park, I saw in the beams of my headlights a leopard suddenly rush out of the darkness and catch a mongoose. I stopped my vehicle to watch. With its back turned to me, the cat quickly ate the mongoose and disappeared back into the dark.

It is remarkable to watch a leopard hunting a gazelle or an impala, its most frequent prey on the African plains. It can also be tedious if you watch with a camera in hand, waiting for the cat to explode into action. The leopard crouches and waits with seemingly endless patience. When the prey's head is down or turned away, the cat seems to glide across the earth, just inches at a time. When the target raises its head, the cat freezes. Close seems never quite close enough for this hunter. Then, as suddenly as lightning, the leopard makes its rush and the hunt is over, often successfully.

As soon as the prey is dead from a combination of shock and the suffocating jaws on its throat, the leopard usually makes that next move unique to its species, dragging the limp carcass into the nearest tree. Male leopards may begin tearing at the meat immediately. Females with old enough cubs will fetch the young to share a family meal. Or, the leopard may crouch and fall asleep on a limb nearby until nightfall before it begins feeding.

There is good reason for immediately caching kills in trees. No kill goes unnoticed, especially out on the open African savanna. Vultures soaring high overhead take note. Attracted by the sound or smell of the gathering vultures, a hungry and much-stronger lion never hesitates to take away a leopard's kill. Then there are the spotted hyenas. A leopard might be able to defend its kill from a hyena or even two of them, but it stands no chance when the whole pack arrives. But stashed high in a tree, the prize is usually safe. However, once on Tanzania's Serengeti Plain, I watched an agile lioness climb just high enough to reach the dangling leg of an impala hanging high over an acacia tree branch. The leopard guarding it clawed at the climbing lion's face but couldn't save its trophy.

There is also new evidence that primitive humans may have scavenged much of their meat from leopard kills. Although studies of the wear pattern on the teeth of early humans reveal that they ate mostly plant foods, there is also evidence that they ate some flesh. Archaeological research from Olduvai Gorge in Tanzania and elsewhere suggests that most, if not all, of their meat was obtained by scavenging rather than hunting. Some scientists now believe that although some meat may have been leftovers from larger predators' kills, most of it probably was taken from smaller hunters such as leopards. Early humans were good climbers and could easily scale any trees where leopards cached their kills.

Although we are unable to watch the behavior of our ancestors, daily interactions between leopards and other primates are revealing. African leopards often prey on baboons while they sleep at night in trees or caves. But during the day, troops of baboons can turn the tables, attacking, and in more than one known instance, even killing a cat. While on patrol, a Tanzanian park ranger recently saw a baboon troop come upon a leopard with an impala carcass in a tree. The large male baboons threatened the leopard from all sides and eventually chased it almost a half mile away. Females and young apes stayed with the leopard's carcass and began to eat until the males returned to take possession.

Leopards also feed on chimpanzees, which in turn will steal tree-stored leopard kills whenever they can. Researchers in Tanzania saw a chimpanzee group surround a leopard lair. A large male chimp then entered the lair and emerged dragging a leopard cub, without reprisal from the outnumbered female still inside. So if you accept the genetic links among all primates, including humans, a predatory/parasitic relationship between ancient humans and leopards is plausible.

It is impossible to write about leopards without describing one most remarkable female cat that, during the 1990s, prowled Kenya's Masai Mara National Reserve. Probably no single wild cat has ever revealed so much to both biologists and eco-tourists so openly for so long. Named Halftail, this leopard was a unique beast.

In 1989, the young female suddenly appeared in a scenic area of huge trees around Fig Tree Ridge and Leopard Gorge in Masai Mara, where Thomson's gazelles, hyraxes, warthogs, and hares—all good leopard fare—abounded. None of the veteran tourist guides and drivers knew the cat's origin, but they estimated her age at about two years. They could not believe how unbelievably tame and habituated this leopard became. At times, she even strolled up to safari vehicles and, while hardly even noticing the occupants, spray-marked tires with her own scent. She also dozed occasionally in the shade of safari cars and used them as concealment in ambushing prey. Word of this unique cat spread, and travelers came to the Mara from overseas just to see her. Peggy and I were among the photographers who together shot more pictures of this one wild cat than any other before or since. But her trusting character was not all that set this leopard apart.

In 1992, Halftail lost a third of her tail in some sort of fight, although no one was nearby to witness it. It might have been a territorial bout with another leopard, a losing encounter with one of the lions then in the vicinity, or most likely, the work of baboons. Male baboons are genuinely formidable animals, and a group of them seldom hesitates to tackle any leopard they catch in the open. However it happened, that's how this female got the name Halftail.

Late in 1992, Halftail had three cubs in a rock den at Fig Tree Ridge. Lions killed one and perhaps the second. But the third, a female soon named Beauty, survived and was raised successfully by Halftail, despite being almost daily surrounded by tourist vehicles. Peggy and I arrived when Beauty was about three months old, and during six weeks of tracking mother and daughter, we learned first hand and close up how a young leopard lives and grows.

Most small lion and tiger cubs have siblings and parents to romp with, but at least during daytimes, Beauty lived a solitary life. She was a fine climber and spent long hours sitting in a tree fork. Then she would suddenly become bored with inactivity and drop to the ground to play a variety of solo games: biting sticks, leaping against limber plants, rolling downhill, batting pebbles in a dry stream bed with a paw, or stalking balls of elephant dung. On those seemingly rare occasions when Halftail returned, Beauty leapt all over her. Most of the time Halftail tolerated, even cooperated with her cub, and the two wrestled or played hide and seek. The play became more vigorous as the days passed.

One morning, we saw Halftail carrying an impala calf kill to where Beauty watched from ambush. Halftail dropped the impala in front of her cub and continued to a shady spot where she waited for Beauty to drag it the rest of the way. Then the two ate nearly all of it. Our African driver-guide, Joseph Gichanga, believed this was the first step in the cub's weaning process. Beauty's childhood experience and training were successful. She was only thirteen months old and hunting on her own when Halftail gave birth to another litter. Normally, an African leopard has new litters at eighteen- to twenty-four-month intervals. Halftail had a fourth litter of cubs in 1996, but only one, another female, survived.

As they become adults, young leopards typically leave their mothers in order to establish their own home ranges, a behavior called dispersal. Biologists are realizing that some mammals, especially newly independent female leopards, stake out a range within or right next to their mother's territory, behavior known as philopatry. This can be either harmonious or cause conflict between the generations.

In 1997, Halftail barely escaped with her life. Either she or a male leopard believed to be courting her killed a goat, and a Masai herdsmen shot her with an arrow. The arrow entered a nostril and lodged in her palate, leaving her unable to eat or drink. Fortunately, she was spotted in this desperate condition by eco-tour drivers who trailed her until veteri-

narians from the Kenya Wildlife Service flew to the scene. Halftail was darted, tranquilized, and the arrowhead was removed by pulling it back through her mouth.

The cat had a fifth litter in Leopard Gorge, but these cubs were killed by an invading male. Next, one of her own older cubs forced Halftail to leave her old territory.

In July 1999, word spread that Halftail had been caught and killed in a Masai snare because she had begun killing their goats and sheep. It was a tragic end for an animal that provided so many with such a rare opportunity to see a leopard living wild and free.

The expected lifespan of a leopard is ten years. At Londolozi and other private game reserves in South Africa where the animals live free of many dangers that face others in the wild, a few female leopards have lived to fifteen years. Halftail lived to be twelve and bore fifteen cubs, of which three were still living in 2001. Twelve years is especially noteworthy in a region such as the Mara Reserve where there is intense competition from lions, hyenas, and especially Masai tribesmen with their expanding livestock herds.

Compared to African leopards, the life history and even the exact range boundaries of the surviving Asian leopards are only vaguely known. In-depth studies of these subspecies of *Panthera pardus* are lacking from India and southeastern Asia. Sri Lanka is an exception.

During a fourteen-month survey of animal life in Wilpattu National Park, where I saw my own first leopards, John Eisenberg and members of a Smithsonian Institution research team reached a number of interesting conclusions

Of all the world's large cats, the leopard appears best qualified to endure on an increasingly developed, plundered planet.

about *Panthera pardus kotiya*, the Sri Lankan cat. There were eighteen leopards resident in the park, the largest carnivores living there or in the entire country. The average hunting territory per cat was about 14 square miles (36 sq km), but those few living where prey was most concentrated roamed over only 2 or 3 square miles (5–8 sq km). Spotted or axis deer account for almost half a Wilpattu leopard's diet by weight, but the cat also kills sambar deer, mouse deer, barking deer, wild swine, jackals, gray langur monkeys, and even the occasional porcupine. Young swine are the most frequently killed prey. But because the leopards made most kills around water holes, and there was little competition from other carnivores, Eisenberg did not find that a kill was ever carried into a tree, as in Africa.

Sri Lanka's Wilpattu leopards stalk langurs on the primates' frequent, dry-season trips to drink at forest pools, as well as in their tree roosts. The sight of a spotted cat just beneath a tree where a troop has sought refuge seems to provoke hysteria in the crown. The monkeys whoop, chatter, and scream. And just often enough, a hyperactive subadult or two will lose its grip and fall to make a leopard's attention worthwhile.

Leopards with black coats, which are simply black color phases of the spotted cats, somehow are far more common in captivity than in the wild.

In July 1993, biologist Maurice Hornocker received word in Idaho that one of his colleagues, Jack Whitman, and a team of fellow Russian biologists had succeeded for the first time ever in capturing, fitting with a radio collar, and releasing a wild Amur leopard, *Panthera pardus orientalis*. This cat, which they called Svetlana, was one of only thirty to forty of this Asian subspecies still clinging to existence in the wild.

Svetlana was found on the eastern edge of Siberia, a region relatively unfamiliar to the Western world. Despite the recent invasion of Siberia by the extraction industries, Whitman, Hornocker, and top Russian Amur leopard experts Dimitry Pikunov and Victor Korkishko had selected an undisturbed study area in the Kedrovia Pad Reserve, just across a bay and 25 miles (40 km) south of Vladivostok.

Tracking the collared cats both by radio and by following their pawprints in the snow, the scientists collected information for the first time about this beautiful subspecies. Amur leopards did not carry their kills up into trees, but usually dragged them away from kill sites to more-secure hiding places. The researchers found where Svetlana and her cub dug a raccoon dog (a relative of the American raccoon) from the snow beneath a boulder and feasted on it.

During the 1980s, Russian government-sponsored deer farms were established to stimulate the depressed local economy. Herds of sika deer were raised on thousands of acres of land enclosed by 10-foot-high (3-m) fences. Each year, the deers' antlers are sawed off and ground into powder that is sold to Asian pharmacies. But those fences were not high enough to keep out leopards, and any deer killers caught inside or outside of the enclosures were shot. No one really knows how many cats were and still are being killed in this way.

In the Eastern Hemisphere, the leopard is the most adaptable of all naturally occurring large land mammals except for people. But it is in the boreal forests, the taiga of Russia, that the adaptation is most intriguing. In many ways, the Amur leopard is like the snow leopard. Both have long, handsome fur coats. And both must conserve energy and body heat during long and brutal winters. By 2001, the Amur had been driven into a final refuge in the Kedrovia Pad Reserve, much like the Florida panther has in the Everglades in the United States. And that last range shrinks every day. The object of Hornocker's multinational program was to somehow save the Siberian, or Amur, leopard from extinction.

Despite government indifference and some local opposition, the program-study has had some success. That first radio-collared female gave birth, and the biologists were able to track mother and her single cub for ten months. The biologists also captured a second animal, a male they called Ooglati. This leopard spent more time outside the boundaries of Kedrovia Pad and wandered too far too often toward human settlements. That could be good or bad news: good because it might reveal an ability to coexist with a human population; bad because of the increased odds of it getting into trouble. While they are optimistic, it is too soon for the Amur leopard research team to make any long-term predictions.

As recently as 2001, many scientists believed that the small leopards known as *Panthera pardus nimr* that were once widespread on the Arabian peninsula were extinct. But by using camera traps, biologist Andrew Spalton discovered small populations still stalking the remote, dry mountain ranges of Yemen and Oman. This proves that of all the big cats anywhere in the world, cunning and adaptable *Panthera pardus* just may be the greatest survivor of all.

Left: *Leopards are the most athletic of the world's great cats—and are certainly the best tree climbers.*

Above: *African leopard country: Generations of the spotted cats have been born and have lived here at Leopard Gorge in Kenya's Masai Mara Game Reserve.*

A female leopard, with cubs hidden nearby, stalks over a green savanna toward an impala fawn.

Left: *The abundant Hanuman langur monkey is a common prey species of the Indian or Asian leopard wherever the two coexist.*

Below: *The warthog is easy prey for all of Africa's larger predators and especially for leopards and cheetahs, which catch the young ones.*

Bottom: *Nursery groups of impala fawns such as this provide easy prey for leopards and cheetahs with their own young to feed.*

A leopard stalks its prey.

An early lesson in survival, a leopard cub drags an impala killed by its mother to a tree for safekeeping from rival predators.

Opposite page: *Except in a few national parks and reserves of India and Africa, leopards are inconspicuous, difficult to see, and even harder to approach.*

Above: *A sudden, drenching downpour foils the African leopard Halftail during a morning stalk of impalas grazing nearby.*

Opposite page: *The near-legendary African leopard, Halftail, once well known to Kenya safari travelers, is greeted by her cub waiting in Leopard Gorge.*

Above: *Asian leopards often follow mountain waterways to explore their territories as well as to hunt, especially in drier seasons.*

Above: *Leopards still share some moun-tainous range in western China with the endangered giant panda. They also sometimes prey on young pandas.*

Right: *The red panda of the Asian moun-tains is another creature on the long list of leopard prey species.*

Opposite page: *From time to time both Asian and African leopards have become man-eaters, a few of them famous far from where they lived.*

Top: *A medium-sized cat, the clouded leopard,* Neofelis nebulosa, *shares some range in southeastern Asia with and is sometimes mistaken for the larger Asian leopard.*

Above left: *The spotted serval,* Leptailurus serval, *of eastern and southern Africa has often been misidentified as a small or young leopard, or as a cheetah.*

Above right: *The caracal or African lynx,* Felis caracal, *shares much of the leopard's range in Africa, but is not often observed.*

Black phase leopards are most frequently found in Malaysia, rarely in their Asia range, and almost never in Africa.

CHAPTER 3

CHEETAHS

Born to Run

The cheetah is born to run. The fastest mammel on Earth, cheetahs have been reliably clocked at speeds of up to 65 miles per hour (104 km/h). Yet in physique, this big cat, *Acinonyx jubatus,* more resembles a dog than its fellow big cats. Its gaunt look, deep chest, and narrow waist are reminiscent of a greyhound or borzoi, and its long legs end in blunt claws. But the cheetah's feline face and short jaws betray it as a cat.

It is this unique anatomy that makes the cheetah's unequaled speed possible. The cheetah's femur bones are longer than those of other big cats, giving it rangier legs. Its spine is elongated and extremely flexible, arching dramatically as the cat runs, thus contracting and expanding almost like a powerful spring to provide the cat a greater reach to each stride. Its short claws are good for traction and for matching the sudden changes in direction of a gazelle running for its life. The cheetah's heart, lungs, and nasal openings are all large relative to its size, which is necessary to supply instantly the oxygen and energy for acceleration in a short, wild chase. The running cat's long, flowing tail gives it balance in any kind of maneuver. At full speed, a cheetah seems to fly over the ground.

Male cheetahs can weigh up to 130 pounds (58.5 kg), stand almost 3 feet (90 cm) at the shoulder, and measure up to 7 feet (210 cm) from their nose to the tip of their tail. Females are a little smaller in all dimensions. These sizes are average for the cats throughout their range, which is limited to open grasslands and woodland savannas of eastern and especially southern Africa. They have roamed, and may still roam, in open highlands, such as up to about 3,000 feet (900 m) elevation in Ethiopia, but never in the humid forests elsewhere on the continent. A few probably cling to existence along the southern fringes of the Sahara Desert where people are few. But any Saharan cheetahs are often pursued by Touareg nomads, who hunt them with dogs whenever they come across their tracks in the sand.

Cheetahs have long intrigued humans. For at least 5,000 years, until fairly recent times, the rangy cats have been kept as pets and especially for hunting. Potentates on three continents boasted stables of cheetahs, which were status symbols akin to harems of concubines, workforces of slaves, and thrones of gold. Sumerian rulers, Egyptian pharaohs, and ancient Assyrian kings had cheetahs for ostentatious display as well as for the chase. In a burial mound dating to 2500 B.C., archaeologists in the Caucasus found a silver vase on which appeared an image of a cheetah wearing a collar. The cats were prized by emperors of imperial Rome and may have appeared in the bloody Colosseum contests. Both Charlemagne and William the Conqueror owned pet

A young cheetah makes its first hesitant steps on its first hunting trip alone.

cheetahs, and Russian princes hunted with cheetahs in the eleventh century. Marco Polo wrote that Kublai Khan kept as many as 1,000 captive cheetahs at a summer palace in Karakorum. The cats were leashed and hooded like falcons until a prey animal was spotted, then released to run their short, successful chase. In the seventeenth century, Emperor Leopold I of Hungary hunted deer with the big cats in the Vienna Woods. Lions and cheetahs roamed the palace grounds of Haile Selassie, the last emperor of Ethiopia.

Female cheetahs two years and older produce litters of up to nine cubs, but seldom more than three or four, after ninety to one hundred days gestation. Cheetahs breed at any time of year, but the peak of births seems to take place as the annual rainy season ends, which is winter in East Africa. Males are able to breed when only about a year old, but do not often have the opportunity until much later. Females can be prolific in giving birth to cubs; like lions, they are able to breed soon after losing a young litter, which is not unusual, or immediately after their cubs depart to fend for themselves. Some cubs can survive independently at about twelve months while others need twice that long before venturing out on their own.

Female cheetahs are almost always solitary, except when being courted or when caring for young. The females do not defend territories, a major difference from other big cats. Instead, they travel far over vast areas of up to 600 square miles (1,560 sq km). Male cheetahs, often in groups of two or three, establish, patrol, and defend territories of 20 to 25 square miles (52–65 sq km); these territories are somewhat greater in average size in eastern than in southern Africa. Single males are occasionally found leading lives more nomadic than territorial.

In recent years, many studies have been done on cheetahs in Africa. Most of these were conducted in larger national parks or within smaller, private, fenced game reserves, and the focus was on behavior in these particular study areas

A dark and rare color variation, known as the "king cheetah," at right, occasionally is seen in southern Africa.

where conflict with humans was minimal. One surprising conclusion was that cheetahs had difficulty maintaining viable numbers in both national parks and private reserves because of the area size restrictions, plus the constant pressure and competition from larger predators. Although efficient hunters, cheetahs are subordinate to lions, hyenas, and leopards, which all often steal their food and kill cheetah young.

The largest population and highest density of free-ranging cheetahs today exists in Namibia, in extreme southwestern Africa. While some of the Namibian cats roam in the vast, dry Etosha National Park, 95 percent of the country's population lives outside the security of park boundaries. In 1991, biologists Laurie Marker-Kraus and Daniel Kraus organized the Cheetah Conservation Fund, a foundation based in Namibia, to protect the country's cats. They were intrigued that Namibia is the only country in the world to include protection of the environment and native wildlife in its constitution.

Surprisingly, the largest number of cheetahs are still found in one contiguous area of 115,000 square miles (299,000 sq km) of Namibia devoted almost entirely to the commercial raising of cattle, sheep, and goats, and where most of the other large predators have been eliminated. As cheetahs can kill small stock and calves, most landowners would like to see the cats removed as well. Between 1980 and 1991, 6,782 free-ranging cheetahs were either shot or live-trapped and moved elsewhere. But the unofficial toll was surely higher than that: Like wolves and coyotes in the United States, many cheetahs are shot and never reported. The fate of Namibia's cheetahs is in the hands of about 1,000 farmers.

Since the Kraus team's arrival, their program has combined sociology, conservation, education, animal husbandry, and wildlife management, all aimed at making Namibians more tolerant of their wildlife heritage, with an emphasis on the cheetahs. The Krauses found that livestock losses could be reduced to nearly nothing by running a few female donkeys with calving herds of cattle. Donkeys are irritable, aggressive, and drive away jackals as well as cheetahs. The biologists also encouraged farmers to try livestock guard dogs and have obtained dogs for the purpose.

By radio-collaring and ear-tagging captured big cats, the Krauses soon learned that cheetahs in Namibia were more wide-ranging than those studied elsewhere. A female might wander over an area of almost 500 square miles (1,300 sq km) that overlapped the areas of several other females.

The researchers also learned about every female cheetah's strong drive to visit and use "playtrees" during their regular travel circuits, behavior that has only been noted in Namibia. These playtrees—mostly old camel thorn trees that Namibian farmers call "newspaper trees"—usually have gently sloping trunks and large horizontal limbs that an animal with nonretractable claws can climb. Cheetahs can run up the trunks and either rest or scan the landscape from a platform above ground. They urinate on the trunk and defecate on the limbs to mark each visit. Long ago, farmers learned that playtrees were the best places to most easily capture the cats. Wire live-trap cages were placed at the bases of playtrees and thorn barriers were located so cheetahs had to enter the cages in order to reach the trunks. Often the vocalizations of one trapped cheetah would summon others, which were also caught.

If a captured cheetah could be sold to a zoo or private game park, it was carted away alive. Otherwise, it was shot. But there was no discrimination between problem animals, such as cattle killers, and others. Survey information by the Krauses showed that playtree trapping tended to upset cheetah travel throughout the area. It could even attract more cheetahs to vacant territories and thereby increase the trappers' problems.

The best news, however, is that attitudes seem to be changing in Namibia toward this unique cat with its long limbs, great speed, and enduring grip on the human imagination.

Cheetahs must rely on blinding speed and surprise to capture prey.

In some ways the world treasure called Serengeti National Park in Tanzania is similar in landscape to Namibia's cheetah country. Serengeti is largely plains—grassland mixed with open woodland—and home to incredible numbers of hoofed creatures, many migratory. Despite the prey abundance, cheetah numbers are lower here and kept that way by constant competition from larger and stronger predators. But if there are any problems here with humans it comes from admirers in safari cars who may inadvertently interrupt the cheetahs' hunting forays, mostly during daytime. The sight of a cheetah strolling across the Serengeti at golden dawn or dusk becomes a memory that no visitor ever forgets.

Another dedicated husband-and-wife team, George and Lory Frame, have long lived in the Serengeti, following individual study cats day and night, sometimes for weeks at a time, making observation notes, and taking pictures capturing behavior. The primary purpose of their work was to evaluate the cheetah's status in the Serengeti ecosystem and, hopefully, to learn why the cats were not more numerous in such a wilderness paradise. That meant carefully monitoring cub birth and survival.

In five years, the Frames found ninety-one different litters, born year around, although most were whelped in the rainy season. The largest litter contained six cubs. Average litter size was four while still in their lair or birthplace. By the time the small, dark cubs were traveling with their mother at from five to twelve weeks old, the average litter was down to two or three cubs, an indication of their vulnerability.

The Frames recorded 1,260 sightings of either individual cheetahs or of groups. Forty percent of these were adult females with cubs and about 30 percent were lone adults of either sex. The other sightings were of adult male groups or of littermates separated from their mothers. All told, the cheetahs' behavior is unlike the social structure of the lions and leopards that share the Serengeti.

Although females lived alone or with their own cubs, the researchers once encountered a group of nine animals, including two adult females that were probably related and seven cubs. Almost all Serengeti males lived in bachelor groups of two, three, or four cats.

The Frames were able to monitor many litters from birth to adulthood. Cubs remained with mothers until twelve to twenty months old, then they gradually drifted away. Separated brothers and sisters usually stayed together for a few more months until sexual maturity at seventeen to twenty-three months. Females coming into estrus would always leave littermates, or perhaps the siblings separated to avoid aggression from the territorial males soon to appear.

The social life of Serengeti male cheetahs could be violent. Bachelor groups (which were probably always brothers) defended well-defined territories of mostly savanna vegetation with acacia woodland and scattered rock *kopjes* frequently marked with urine, feces, and claw scratchings. The males left territories only briefly to hunt nearby or to drink. For as long as four years, bachelor bands maintained their territories until killed or driven away by a stronger coalition. Only a few solo males survived precariously as nomads without territories, and they were always furtive cats. Once, the Frames witnessed a territorial clash that ended with one invader dead, another male badly injured, and a third flushed out. The worst injury to the defenders seemed to be bloody faces.

Male territoriality regulates—and usually limits—cheetah populations. As stronger males claim more habitat for territories, there is more conflict and death. Increasing numbers of the cats are forced into marginal areas where people,

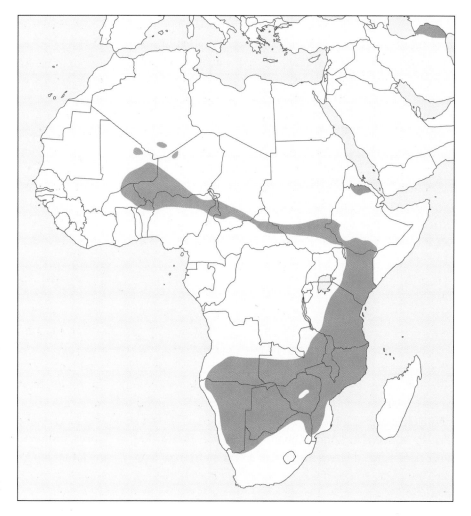

Above: *Track of the cheetah,* Acinonyx jubatus

Front adult paw length: approximately 3¼ inches (8 cm)

Right: *Range of the cheetah,* Acinonyx jubatus

other carnivores, and a smaller supply of prey create additional hazards. Also, too many territorial males can unduly harass females trying to raise families. The Serengeti studies revealed that while equal numbers of male and female cubs were born there, the total population contained twice as many adult females as males.

Every cheetah's survival depends on its hunting skill. Of all the great cats, these daytime hunters are more active in early mornings and least active during hot afternoons. The Frames recorded 493 hunts in which the cats managed 203 kills. More than half the prey were Thomson's gazelles.

Most successful hunts are the result of silent stalks lasting from a few minutes to more than an hour. The cats seem to select the most-vulnerable, least-alert targets, such as a single gazelle or one of a small group, which has fewer keen eyes watching out for danger. Stalking enables the cheetahs to approach within 30 to 150 feet (9–45 m) of a mammal almost as fast yet more long-winded than itself. At the instant the hunter launches an attack and is spotted, the startled target may waste a fatal second or two before darting away. What follows is a race beautiful and breathtaking to watch.

Cheetahs are sprinters rather than long-distance runners, however, and though spectacular, most chases are brief, only a few hundred yards at most. Successful ones end when the cat pulls close behind or alongside its target and, with a front foot, trips the prey. Often a cloud of dust blurs the flying feet and bodies as the winner secures a stranglehold on the loser's throat. Just as often, no contact at all is made, and the prey leaves an exhausted cat behind, limp on the ground and panting hard.

During seasons when an abundance of baby antelope are born, a cheetah's life may be a lot easier. It can hunt by simply walking and watching for crouched fawns.

There also have been scattered reports of cooperative hunting among cheetahs. One cheetah stalks and then flushes an antelope directly toward a partner, who is crouched in ambush. Cheetahs have also been seen hunting in relays, where a second or third cat takes up a chase after the original stalker is drained of energy. Larger animals such as wildebeest and topis might be taken in this way.

Among the great rewards of wildlife research is the discovery of exciting new information or of exposing popular misconceptions. The Frames were surprised to find how unsuited the cheetahs are to living and hunting in short or overgrazed grasslands. Instead, the cats need higher cover both for stalking and for concealment from other, stronger hunters. Except during long, dry spells, most of the Serengeti today is suitable cheetah range most of the time.

* * *

Some cheetah researchers are concerned about the number of incidents of aberrant cheetah behavior reported in scattered parts of their range. Cannibalism is a notable example. Tourists in South Africa's Kruger National Park watched a mother grasp each of two cubs by the neck and shake them until dead. She then ate one before walking away from the scene.

Luke Hunter, who has studied cheetahs in South Africa since 1992, observed and photographed cheetahs eating cheetahs in the Phinda Resource Reserve on two occasions. A pair of males that had been living normally in the reserve for fifteen months killed an intruder and then proceeded to maul the carcass for an hour after it was dead. They then dined on it. A year later, the same two males caught and ate another intruder. One explanation for the behavior might be that the two cannibals had been captured elsewhere and re-introduced here in the northern KwaZulu-Natal, an area rich in antelope prey. No wonder biologists are baffled.

In 1926, scientists in England were mystified by photographs they had received of a strangely patterned animal skin sent from Rhodesia (now Zimbabwe) by a Major A. L. Cooper. Solid, dark stripes ran along the back of the skin, and irregular dark blotches decorated the rest of it. At the British Museum in London, mammal division curator R. J. Pocock concluded it was the skin of a mutant form of leopard. But after examining the skin himself, Pocock changed his mind and declared it was a new, rare cheetah species, which he named *Acinonyx rex*, or "king cheetah." That assumption was also wrong.

Since then, only a few living specimens of the cats bearing the king cheetah pelts have been sighted in northern Transvaal, Zimbabwe, and Botswana. Thus it was no wonder a great stir was created in wildlife circles when, in 1981, a cub from a litter of five commonly spotted siblings was born with a king cheetah coat at the De Wildt Cheetah Centre, just west of Pretoria, South Africa.

The De Wildt Cheetah Centre was established in 1971 in the Magaliesberg foothills by animal enthusiasts Ann and Godfrey Van Dyk at a time when cheetahs were not faring well in the republic. Despite the fact that cheetahs never before bred well in captivity, the Van Dyks succeeded in whelping hundreds for further study, for zoos, and for restocking the wild population. Among these were a dozen of the varicolored "kings" that still are seldom if ever reported roaming free.

The cheetah is born to run, and today its entire existence is a race for life. Cheetah numbers are difficult to estimate, and an accurate census is not possible. In 1900, about 100,000 were believed living in forty-four countries of Africa and Asia. Since then, the species has been completely extirpated from twenty countries. Best estimates in 2001 were that from 4,000 to 12,000 cheetahs remained in mostly scattered pockets within twenty-four African countries and, just possibly, in Iran.

During the 1979 revolution, Iran's numerous nature reserves were invaded, villages built, and the wildlife wiped out; the reserves were unstaffed and poachers shot all game in sight. Recently, Iran's Department of Environment has worked to rehabilitate what little habitat is left. Using camera traps, biologists estimate that a few cheetahs, *Acinonyx jubatus venaticus*, still roamed northern Iran in 2002 and sought international help to save them. Nevertheless, the future of the cheetah in Iran does not seem bright.

Cheetahs are not faring well in Africa either. Poaching has taken a toll and predator control a much greater number, but these are not nearly as damaging as the loss of habitat across southern Africa from Nairobi to Capetown. The cheetah is definitely an endangered species and from a cause far worse than random killing and habitat loss.

A much greater threat to the species may be genetic. Today, cheetahs are critically short of a variety of genes. The grim news comes from experiments by a team of the National Cancer Institute led by geneticist S. J. O'Brien. Blood tests on cheetahs scattered across southern Africa, where they are most numerous, revealed a low genetic diversity. In other words, it showed that there had been far too much inbreeding for far too long. Whatever the number of surviving cheetahs in 2001, their gene pool is small, much smaller than it should be for the existing number of animals. It is probably too small to carry them through some sudden adversity such as a serious epidemic that could kill all instead of some of the population. With such a limited gene pool, none of the surviving cheetahs may carry a trait that would allow them to overcome such adversity: the ability to reproduce. Male cheetahs tested recently have an alarmingly low sperm count and too many abnormally shaped sperm, symptoms that are also characteristic of inbred livestock and inbred laboratory mice.

Cheetahs are approaching, or have already reached, a blind alley or bottleneck that humans cannot soon rectify. All we can do is give the cats time and space. Although a depleted animal population can recover quickly, a depleted gene pool cannot. Hundreds, maybe thousands, of generations must be born and reproduce to reach a safe level of genetic variation.

Cheetahs might recover. The American bison came close to extinction a century ago, but is reestablished now and considered safe. Populations of elephant seals and fur seals in the Southern Hemisphere were close to zero after decades of commercial hunting, but their numbers have rebounded, although with a severe deficit of genetic diversity. The great mystery with the cheetahs is what happened to the missing genes. A greater mystery is whether their species can endure without them. Our hope is that this endangered species will survive and still gracefully stalk its environment for a long, long time.

Above: *A morning squall clears the atmosphere over the Mara Plain in Kenya, cheetah country.*

Right: *A mother and young cheetah use a termite mound to scan the surrounding Serengeti Plain in Tanzania for game.*

All photos: *The long stride of a cheetah running at full speed is an exciting spectacle to watch—and one that ends all too quickly. A cheetah's long legs give it unmatched speed during a short chase. But this handsome hunter is seldom successful during a long pursuit.*

A cheetah surveys the African savanna for prey.

Top: *A cheetah keeps its eye on this Thomson's gazelle with a newborn fawn, which will make an easy meal.*

Above: *The longer the chase, the better the chances for the impala's escape as the cheetah tires early.*

A cheetah carries a gazelle fawn, still alive, to be eaten out of sight in a shady place.

Above: *A short race and this cheetah snatched a young warthog from a family running at full speed to escape.*

Left: *A family of South African cheetahs shares a springbok kill that has been dragged into a shady place.*

Opposite page: *Beautiful landscapes such as this in Kenya's Samburu National Park are home to all the region's great cats and an abundance of their natural prey.*

Above and left: *Sibling sociability or bonding is often a ritual after a successful hunt, when stomachs are full.*

Above: *All cheetah family members intently watch a herd of antelope passing in the distance. The mother will decide whether to pursue or not.*

Opposite page: *A young cheetah of about four months plays with a stick while its mother is away hunting.*

Scattered trees on the open savanna serve as shady resting places for cheetahs. In southern Africa, they are often called "cheetah trees."

Heavy rains produce a lush grass cover on the plains, a situation not ideal for this cheetah beginning a morning hunt.

CHAPTER 4

TIGERS

The Regal Striped Cat

The soft Indian twilight was fading into dusk when I sensed that something was moving. All at once the forest fell quiet. No birds sang, and even the insects seemed to stop whining around my ears. From my confinement in a tree blind, I could dimly see the carcass of a buffalo on the ground below, but nothing else. Buck fever made me forget the cramped muscles in my legs.

Time passed in slow motion. Eventually the buck fever dissipated, and I prepared to leave my perch. But it wasn't until I capped my telephoto lens that I saw the tiger. It appeared beside the bait below as silently and as suddenly as a distant bolt of lightning. A moment later, the cat was tearing flesh from the buffalo's rump.

Maybe the tiger heard my heart and pulse pounding, for seconds after beginning to feed, it looked upward, directly at me. It hesitated and then dissolved into the dusk. Daylight was gone before I felt calm enough to climb down from the blind, a few years older.

Though brief, that long-ago first encounter with a wild tiger in India remains unforgettable. It was my premiere glimpse in the wild of one of the most splendid animals on earth. Since then, I have concluded that no first encounter with a tiger, even in a circus or a zoo, is ever really forgotten.

The tiger, *Panthera tigris*, is a native of one continent, Asia, and is the only one of the big cats that is striped. Though the tiger's striking pelt of black stripes on orange makes it the most readily recognizable animal in the world, the striping patterns of individual cats are as distinctive as the fingerprints of humans. No two are the same.

Powerful, majestic, ruthless, exquisite, enigmatic, and mysterious are a few of the adjectives commonly used to describe tigers. Depending on where they live today, tigers are despised or worshipped, held up as symbols for saving the world's environment or as a valuable commodity to be killed and sold for its pelt, penis, and other body parts. Athletic teams and military units all over the world use the tiger as an icon of their prowess. Advertisers use the striped cat to sell everything from gasoline to potions promising eternal youth.

Most villagers living on the fringes of tiger country today regard the cats with fear and hate. To them, tiger survival is justified only for tourism value, specifically for tiger watching by wealthy visitors. But the man who probably knew tigers more intimately than anyone had a more courtly, anthropomorphic view of them. "The tiger," naturalist Jim Corbett wrote in 1944 in his best-selling book, *Man-Eaters of the Kumaon*, "is a large-hearted gentleman with boundless courage."

* * *

Bandhavgarh National Park is among the few last refuges of the Indian or Bengal tiger, where this female was photographed from elephant back.

Because of their secretive habits, solitary nature, and scarcity today, tigers have never been studied as widely or as thoroughly as lions. Until the 1960s, virtually everything we knew of the species was culled from myth or observations made over the barrel of a rifle.

Then in 1982, biologist George Schaller went to India to track and scientifically study the Bengal tiger. Under the auspices of Johns Hopkins University, Schaller focused his work on Kanha National Park in Madhya Pradesh state, an unspoiled region of about 100 square miles (260 sq km) in north-central India. Here the land had not been overgrazed by livestock nor had the cats been ruthlessly slaughtered. The necessities of life for tigers—food, water, and cover—existed. Kanha remained one of the most beautiful natural areas in this densely populated land with its forests and meadows, *maidans* (ponds), and an abundance of native species, from peafowl to several types of deer. Schaller realized that tiger behavior would vary from area to area throughout the big cat's wide range, but he believed that Kanha tiger behavior would be fairly typical.

Even with the sophisticated equipment available today, tiger study is neither easy nor convenient. Schaller opted to track the big cat on foot, unarmed, day and night, as much as his strong, young legs and desire allowed. Many moonlit nights were spent watching from tree blinds and filling notebook after notebook with observations. At least once he experienced abject fear when he inadvertently walked within a few feet of a sleeping tiger. When the cat suddenly awoke, the two locked eyes. Schaller suffered a moment of dread, but the tiger growled, and then simply stalked away.

The secretive tigers proved difficult to study for Schaller. However, he soon found that tigers were easy to distinguish from one another by their individual markings above the eyes.

Wherever in the Old World the tiger still clings to existence, it is at the top of the food chain.

He also believed that many of the study animals recognized him in turn by sight as well as smell. Though the cats tended to avoid him, one tigress with four cubs finally allowed him to follow the family, monitor their growth, and document their activities for months.

Searching for food consumes much of the tiger's life. Most hunt alone, walking alertly and silently over familiar paths. A male tiger's territory may vary greatly in size, acting as a spacing mechanism that decreases confrontation or competition between males for females, as well as for the available food supply. The territory of a female, which it marks by urine spraying, is smaller than that of a male, ranging up to 25 square miles (65 sq km). A male's territory usually overlaps the hunting spaces of several females.

Tigers hunt primarily between dusk and dawn when it is cooler, and when deer, wild boar, and other prey feed in open areas where they are easier to stalk from cover. But a really hungry tiger, such as an old or injured one or a female with cubs to feed, will hunt at any time.

A healthy adult tiger is a powerful animal. It has acute senses and is swift afoot over short distances. Its arsenal also includes formidable claws and teeth, with canines as long as a human thumb that would seem to make catching and killing prey an easy matter. But Schaller reported that this consummate killer works hard for its meals.

During one hot, dry season in Kanha, about 800 big-game animals, mostly deer, had concentrated in a 5-square-mile (13-sq-km) area to be near the few water holes that had not dried up. It should have been a banquet for the meat-eaters, nonetheless Schaller noted that several nights of hunting might pass between a tiger's kills.

Conditions must be just right for success because prey species also have keen senses and many can run faster than tigers, given a start. Enough cover is necessary to allow the big cats to stalk within 40 to 60 feet (12–18 m) before making the final surprise rush. Schaller estimated that Kanha tigers made twenty to thirty unsuccessful attempts for every kill. During a number of visits to Ranthambore National Park in northern India, I saw the sambar and spotted deer gather around the shrinking reservoirs during the dry seasons, often grazing on aquatic vegetation in neck-deep water. Tigers also lurked here and stalked what seemed to be easy, exposed targets. But not once from sunrise to sunset did I ever see a tiger score.

To a human observer, a tiger's hunting expeditions appear to be fraught with frustration. Schaller once observed a tiger patiently stalking a herd of deer, only to have a jackal interrupt the big cat's hunt. The jackal barked, exposing the hunter and spooking the deer. The tiger quickly turned on the jackal and chased it away. Resuming her hunting, the cat eventually found a small band of swamp deer, but this time the deer scented the cat, faced her, barked shrilly, and ran. Before long, the tigress found still another group of swamp deer and this time just missed one with her slashing forepaws. Then, as the cat walked away, moaning softly as they sometimes do, all the deer turned and trotted a safe distance behind the cat as if in contempt. They knew that without the element of surprise, no tiger could catch them.

During his studies, Schaller examined the remains of almost 200 tiger-killed deer. Many were very young or old animals. Only a few were prime-of-life specimens, those with the best noses, ears, and eyes. A couple of exceptions were less-agile does heavy with fawns.

Except to tackle the young ones when possible, tigers avoid such large and formidable animals that share their habitat as elephants, rhinos, and gaurs, which are wild cattle that can weigh more than 1,500 pounds (675 kg). Reports exist of wild boars as well as gaurs disemboweling tigers. A truce seems to exist between the cats and sloth bears. Tigers often rob leopards, their main competitors, of their kills and will also eat any leopard cubs they find. But catching a healthy spotted cat is not possible. A tiger has no margin for error when hunting any species of large prey.

During droughts and monsoons or when hoofed prey is scarce, tigers may be reduced to eating anything from jungle fowl and peacocks to frogs, crabs, langur and rhesus monkeys, even grass and wild berries. A hungry tiger will readily consume any carrion it finds, no matter how badly decomposed. In fact, its keen sense of smell can locate a decaying carcass from a great distance, and this ability can lead to a cat's doom as poachers and cattle owners have learned to lace carcasses with poison.

After his studies in Kanha, Schaller concluded that throughout India (and possibly wherever tigers still live) livestock, cattle, and domestic buffalo make up a large part of tiger diets nowadays for two reasons. First, so much of the great cat's natural prey has been reduced or eliminated, often because the vegetation that prey animals require has been decimated by livestock. Second, domestic livestock is much less wary and easier to kill than wild stock.

Maintaining a vigil from a baited blind near a tiger's kill is both tense and revealing. Every cat guards its prize and remains close to it until everything digestible—entrails and some softer bones as well as all of the meat—is consumed. To safeguard their kill, tigers with full bellies will doze in the nearest shade or submerged in some forest pool. Some tigers will drag a carcass to a gully or another place where it is better hidden. Some also cover the meat with leaves and earth. The cats return as often as they become hungry enough to refuel or to drive away such scavengers as jackals, dholes, jungle crows, and vultures. Once in Bandavgarh National Park, I watched a tigress with two small cubs expend as much energy driving away scavengers as she may have used to catch the deer in the first place.

A tiger might linger around a dead spotted deer for two or three days until nothing but bones remains. Or it might stay as long as a week if the kill is a swamp deer weighing as much as 300 pounds (135 kg). Schaller calculated that to maintain prime physical condition, an adult tiger requires 15 to 20 pounds (6.75–9 kg) of meat per day, or 3½ tons (3,150 kg) a year. Considering that less than half of every prey animal is edible, each cat must kill about 4½ tons (4,050 kg) of prey on the hoof annually, or about seventy adult spotted deer or thirty domestic cattle, or some combination of these.

But these estimates can be misleading. Tigers are forced to make even more kills because people often chase them away before they can feed, especially if they have killed livestock. And wherever they are regularly harassed or shot over a meal, they soon learn never to return to it.

Tigers differ from lions in that both sexes hunt, alone, to live. By comparison, male lions are inept or reluctant hunters, because lionesses do most of the hunting for the pride. A tigress may even share a portion of her hunting territory with other females, but males are not so tolerant, especially of other males.

While it is true that tigers hunt alone and are far more solitary than lions, they may be more social than commonly believed. Because tigers hunt so much at night, nocturnal meetings along territory boundaries are inevitable. And tigers are certainly able to communicate with each other and over long distances. Anyone lucky enough to spend still nights in tiger country is almost certain to hear the tigers' deep, "Here I am" roaring over and over, perhaps coming from different directions. Depending on how securely you are located, the roaring can be exciting, haunting, or downright chilling.

As tigers hunt and crisscross their territories, both sexes wave their tails at intervals and spray a mixture of urine and distinctive scent along the trails. It is a lasting and powerful odor—a calling card—that allows the cats to find or avoid one another. Unfortunately, in years past it also allowed skilled woodsmen to track them by scent alone and kill them.

Schaller observed many instances of tiger sociability that other biologists would later also describe. A tigress might roar to call her older cubs to the site of a kill. Or two females with cubs might share a kill for several days in the style of African lions. Schaller also became acquainted with one large male living in his study area that often mingled easily with females. Once, he saw this male share a bullock kill with two tigresses and four cubs. The females made no effort to keep their cubs away from him—which possibly could have been his own—as they romped and climbed all over him. Most females are reluctant to have males come close to their young.

Males mate with any estrus female they find on any territory. Tigresses give birth to one to seven, but usually two or three, cubs about three months after mating. Seldom do more than two cubs survive to one year and beyond. Birth occurs in some inconspicuous place, away from hunting trails, often in a ravine, beneath a deadfall, or in a cave. The bond between mother and cubs is the most lasting of the species. The newborns are blind and helpless for a while, but may be left alone for long periods while the mother hunts. When the

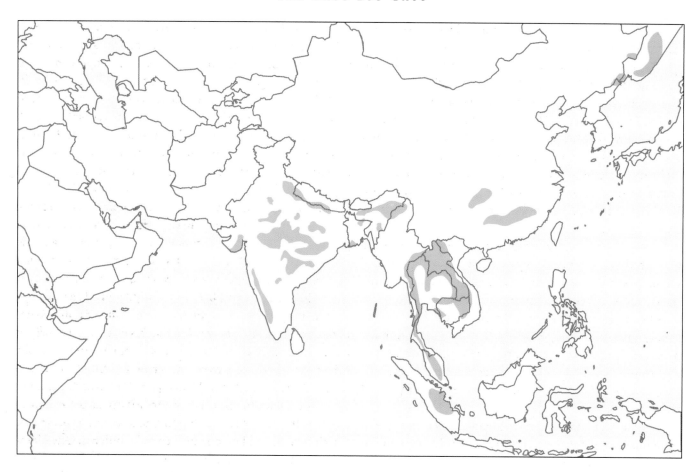

Above: *Range of the tiger,* Panthera tigris

Left: *Track of the tiger,* Panthera tigris
Front adult paw length: approximately 5⅝ inches (14 cm)

cubs are a few weeks old, the mother may begin to bring them meat, and by six or seven weeks, they leave the natal shelter with their mother and thereafter lead a nomadic life. They alternately wait in secluded places along the hunting trail and follow the tigress to her kills. Full independence comes slowly.

When left alone, the cubs develop a pecking order, and the strongest are the first to feed at kills. When hunting is poor or prey is scarce, the weakest ones soon succumb. When times are good, the family gorges. Schaller observed one tigress with four small cubs that consumed all but the larger bones of a gaur calf, a total of about 85 pounds (38 kg), all in one night.

At about one year of age, tiger cubs begin to "help" their mother hunt, but until they have had a good bit of experience, the help is an impediment. By two years of age, cubs are finally able to hunt on their own and seek territories of their own. Finding their own territories is not an easy task, and many cubs slowly perish. Surviving in an ever-decreasing savage jungle, killing to live, and being surrounded by an increasing human population is a formidable challenge. Some tigers become man-eaters because no prey is so easy to catch or so abundant as people walking in a woods.

* * *

With new technology, researchers can today learn much more about the lives of tigers than was previously possible. One of the most useful devices is the radio collar, which allows scientists to track and monitor the actions and behavior of their subjects from miles away. Due to radio tracking, researchers now have growing evidence that some tigers have learned to live near human beings, at times almost among them, without being detected. Biologist Mel Sunquist of the Smithsonian Tiger Project radio-collared many tigers and observed numerous such furtive cats.

One morning, Sunquist stood on a bank of the Rapti River in southern Nepal watching as laughing women washed saris in the warm current and then beat their wash noisily against rocks while naked children splashed nearby. None of the women or children knew that crouched in a thicket 60 feet (18 m) away was a tigress fitted with a radio-collar transmitter that was sending signals to Sunquist's receiver. On another occasion, he listened, horrified, as more than 100 villagers walked along a forest trail almost within touching distance of a radio-collared cat feeding on its kill. During his many years of tiger research, Sunquist often found tigers within pouncing range of people. Such experiences left him viscerally aware of the tiger's ability to coexist with humans. Yet too often after just one incident, one accidental encounter between human and tiger, a person is killed. Then man-eating can become an epidemic.

There is good evidence that only immense pressure to feed a hungry family eventually drives some tigresses to become man-eaters. Woodsman and naturalist Jim Corbett did not buy into that theory. He believed man-eaters adapted to the unnatural diet because of unnatural circumstances and wrote, from his own experience, that nine of ten man-killers suffered from festering or disabling wounds and, in the tenth case, old age. Their wounds may have come from gunshots, fighting injuries, or the imbedded, infected quills of a porcupine.

Perhaps no human will ever understand tigers as well as Corbett who, at times, actually lived like one. The orphaned son of a British civil servant in Naini Tal, a Himalayan hill station, Corbett helped support his mother and eleven siblings by subsistence hunting in the exquisite wild India that is now gone forever. For Corbett, that grave responsibility became a forest idyll. He acquired an encyclopedic knowledge of the birds and sloth bears, the beautiful deer and tigers, and became more comfortable in their environment than among people.

At the beginning of the twentieth century, an expanding human population was clearing more wilderness and tiger country to farm and raise cattle. Conflict with tigers was inevitable, widespread, and probably likely to increase. When every other means to trap or dispose of the marauding killers failed, Corbett was the man recruited to dispatch the culprit. In the end, Corbett destroyed man-eaters that had killed an estimated 1,500 Indians.

A number of the man-eaters Corbett tracked down became famous far beyond the borders of India. One tigress known as the Champawat Man-Eater made 434 known human kills in Nepal and Kumaon during a four-year period. When Corbett finally caught up with the animal, he found that a bullet had smashed both her upper and lower jaws, making it almost impossible for her to kill natural prey.

Another tiger known as the Mohan Man-Eater, which Corbett killed in 1926, had an early encounter with a porcupine. The hunter extracted more than twenty-five quills, some 5 inches (12.5 cm) long, from one leg and the chest of the dead tiger. That leg was virtually useless, and Corbett wrote that the flesh under the skin from the chest to the paw was dark yellow in color and soapy to the touch. The Mohan cat, which was slowly wasting away, had actually moaned in pain as it walked.

Though Corbett had far more than one man's share of close calls with man-eaters—some of which stalked *him*—he had tremendous regard for tigers. He considered them gentle and harmless to man except in those cases when strange circumstances transformed them into man-killers. In the latter years of his life, the great old woodsman believed that tigers were doomed and wrote that they should not be hunted except with a camera.

* * *

Now and then in recent years, a man-eater makes news headlines and is shot, usually along with some innocent cats. But these incidents are declining because so few tigers are left to stray. When India was first "discovered" by Europeans in 1498, there might have been 150,000 tigers stalking the subcontinent. Following World War II, only an estimated 30,000 to 40,000 tigers still roamed free in India. In 2001, the most optimistic estimate by biologists is that only about 3,000 animals survive—and that population is rapidly declining.

Sport hunting seems to be an easy reason for the decline, and recent Indian history might support it. For decades, Indian royalty organized massive hunts, and their guests slaughtered an obscene number of the big cats. In a single hunt, thousands of beaters on foot and hundreds more riding elephants were coordinated to scour vast areas, infantry style, to drive the game past batteries of royal guns. Tiger hunting in that fashion was slaughter, not sport. In 1951, the Maharaja of Surguja modestly noted that his own personal tally of tigers was 1,150. The Maharani of Jaipur revealed that she had shot twenty-five tigers by her twenty-fifth birthday. In the 1960s, a visit to the Maharaja of Wankaner's palace in Gujarat found the princely trophy rooms carpeted wall to wall with tiger pelts. Tiger hunting also became a pastime for rich *nawabs* and bored British colonial officials. One Major Rice wrote home that, during his tour of duty from 1850 to 1854, he had "shot or wounded 150 tigers in Rajasthan state during his spare time." Britain's King George V killed thirty-nine during one orchestrated hunt in 1939.

A Siberian tiger begins a rush toward a band of red deer feeding at the edge of a forest.

Still, the bottom line is clear: The plight of *Panthera tigris tigris*, the Indian or Bengal tiger, cannot be blamed on gunning alone. Today, India's frightening overpopulation is the main reason and one that's nearly impossible to check. Few of the splendid cats can live outside of national parks where they are fortunately easy to see today.

An unknown number of tigers persist in the Sunderbans, literally the "Beautiful Forests," a swampy mangrove habitat where India and Bangladesh adjoin on the Bay of Bengal. These tigers swim between islands to feed on fish, crabs, birds, and the occasional local fisherman or honey hunter.

The Sunderbans has always been an extremely perilous place for honey collectors or anyone else to roam alone. During one ten-year period in the late 1970s and early 1980s, 429 deaths were reported. Authorities made several attempts to drive tigers away with truckloads of free fireworks. It didn't work. Neither did prayer to images of the Hindu goddess of forests, Banbibi, or Shah Jungli, the Muslim king of the jungle.

Eventually the Sunderbans tiger reserve director, Pranabes Sanyal, tried a unique experiment. Mannequins were dressed in discarded clothing to give them a true human smell and then were wired to 12-volt automobile batteries. When grasped by a tiger, the cat received a stinging jolt of electricity. By moving the dummies around wherever tigers seemed most active, rangers hoped to train tigers through avoidance conditioning that people are painful targets. The trick may have helped. Tigers did stalk and attack the fake woodcutters and honey hunters, and some got the shock of their lives. Villagers noted that some tracks led right up to the dummies then suddenly veered away. In many areas the killing stopped, but not entirely.

Another problem spot was Dudhwa National Park, a remnant of original forest in the Himalayan foothills of northern India. Until 1980, Dudwa teemed with malaria and was thus sparsely inhabited. But when dousing with DDT killed the mosquitoes, people arrived en masse, farming and planting sugar cane right up to the park boundaries. Tigers quickly killed 138 farmers in a seven-month period. The problem was mostly solved here by immediately hunting down the culprit cats, but killings still do occur.

Chitwan National Park in southern Nepal protects a hundred tigers or so, and a smaller number probably still stalk the northern Myanmar highlands of Burma. A few tigers travel into southern Bhutan from adjacent Indian sanctuaries in Assam. Recently, tigers have also been reported in the mountain areas in northern Bhutan. In fact, a tiger was captured on film by researcher Pralad Yonzon at 9,840 feet (2,952 m) in Thrumsing La National Park. It is the highest elevation ever recorded for a tiger; this one lived among red pandas and capped langurs.

Asia's other subspecies of tigers have fared even worse than the Bengal tigers, whose threat is strictly from human expansion. Elsewhere in Asia, tigers are still trapped and hunted for the potions, powders, or cremes purported to cure a variety of human ills, and their meat is in immense demand to aid stomach disorders. Consider the following sixteenth-century Muslim treatise that lists the tiger pharmacopoeia: "use brains and tiger 'oil' to cure laziness and pimples; eat the gallstones for better vision; add dried tail to the bath as a skin balm; make pills from the eyes to stop convulsions; mix fat with roses for a face cream that will bring honor to the wearer; wear fangs, claws and whiskers as love charms or amulets to guard against danger." At least not much was wasted. Although the treatise is 400 years old, its advice is still believed valid by many Asians. In fact it is valid enough that, even in troubled economic times, wild tiger meat fetches more than $500 per pound on the black market. Dried tiger penises sell for far more than that.

It is no wonder that during the past half century three tiger subspecies have vanished altogether from the wild. First to go was the Bali tiger, *Panthera tigris balica*, which was declared extinct in 1950. Almost nothing is known about this race.

Next to disappear was *Panthera tigris virgata*, the Caspian tiger, a small-to-medium-sized, dark-colored cat. The Caspian tiger once ranged from northern Iran through Turkmenistan, Uzbekistan, Kyrgyzstan, and Afghanistan to the Aral Sea and Lake Balkhash in Kazakhstan, even as far west as Asia Minor.

From time to time there are rumors that a Javan tiger, *Panthera tigris sondaica*, or two still slinks in a small jungle somewhere in Java, but the subspecies was considered extinct in the 1980s. No one knows when or how the last one died, or where its body parts were sold.

It may not be long before the last roar of the South China tiger, *Panthera tigris amoyensis*, is heard. The South China tiger once roamed throughout southern, central, and eastern China and into Korea. But despite its prominence in Chinese art and literature, for far too long its essence has appeared in everything from powders and balms to wines and love potions sold in shops from Shanghai to Taipei, San Francisco to

A male Siberian tiger scent marks a tree in its territory and tests the air for the aura of a female in season.

Vancouver. An estimated 3,000 to 4,000 big cats were destroyed during China's war on pests and vermin just after World War II. Ten to fifty South China tigers may still exist, but if so, their habitats are being closed in on by Chinese cities and villages that never stop expanding.

Probably more Sumatran tigers, *Panthera tigris sumatrae*, survive in zoos and private collections today then in the remaining wildlands of Sumatra, Indonesia. For a long time after the country became independent of Dutch rule, all animals were hunted without restriction. Now the number of surviving tigers is estimated to be from fewer than 100 up to an unlikely 600. No one really knows.

In 2000, Cambodian researchers, funded by the World Wildlife Fund and the Wildlife Conservation Society, began surveys of the country's once-abundant wildlife. Using automatic, infrared-triggered camera traps along game trails, the researchers exposed some surprising images of leopards, clouded leopards, sunbears, wild dogs, elephants, banteng, and gaurs. Most surprising of all were photographs of a tiger, the first ever taken of the big cat in the wild in Cambodia. This cat was one of the 1,000 to 1,500 of the Indochinese tiger subspecies, *Panthera tigris corbetti*, thought to still live in southeastern Asia. This gave conservationists hope that tigers could also thrive in the new national parks of Vietnam, Thailand, Malaysia, Laos, and Myanmar where there is a new interest in natural-history tourism.

The tiger subspecies with the best chance to survive is the Amur or Siberian tiger, *Panthera tigri altaica*. It is also the largest of the tigers, with the heaviest males weighing more than 500 pounds (225 kg).

If one man is somehow able to revive interest in a species, to retrieve it from the brink of extinction and jump-start its recovery, that person may be Maurice Hornocker and the species is the Siberian tiger. Working with a cadre of dedicated Russian biologists and Howard Quigley of the Hornocker Wildlife Institute, he has practically commuted back and forth from his base in Bozeman, Montana, to Siberia in the cause of tiger preservation. One result is the Sikhote-Alin Biosphere Reserve in Russia, one of four wild sanctuaries especially for tiger habitat.

Hornocker and Quigley bring their great experience with North American mountain lions to their Siberian tiger investigations. Earlier, on the White Sands Missile Range in New Mexico, Hornocker discovered that a single cougar female had acquired a taste for wild sheep and alone was endangering the desert bighorns living on the range. Recently on the Sikhote-Alin Biosphere Reserve, he and Quigley discovered a similar situation. A 400-pound (180-kg) male Siberian tiger had developed an even stranger preference for just one prey: brown bears. Although red deer and other game were readily available, this cat stalked and ate bears almost twice as heavy as itself. The biologists tracked the tiger through the snow to eight separate bear kills, all of which seemed to have been accomplished without great effort, except one. In that

Rarest of all tigers is the Indochinese subspecies, Panthera tigris corbetti. *Only a few hundred may still roam in the remotest Cambodian and Viet Namese jungles.*

kill, there was evidence of a vicious battle with bits of bear hide strewn over a wide area, but the tiger had won. The only conclusion was that, without natural predators, bears forage worry free, with heads down, and are easy to surprise.

At one time, the Siberian tiger subspecies populated a wide swath of cold, gloomy forest across Siberia and Manchuria into China, subsisting mostly on red deer and wild boars. Poaching for both tigers and their prey always existed, but a sharp upswing began with the demise of the Soviet Union and the collapse of government funding for any conservation. At the same time, an increase in road construction for mines and timber cutting, much of it illegal, invaded tiger habitat and made poaching easier. According to Hornocker, the winters of 1992 through 1994 were especially tough on tigers. Lack of law enforcement along Russia's Amur River border with China made it vastly easier to trade tiger parts to where they had the greatest value. All at once, a tiger carcass was worth about $15,000, more than the average Siberian could earn in two or three years of hard work.

Today, Siberian tigers are largely confined to the coastal strip of easternmost Siberia north of Vladivostok. With funding from the National Geographic Society, Hornocker has helped finance, inspire, and encourage his Russian partners to study and vigorously protect the resource that is so important to their country and the world. He has also succeeded in stiffening Russia's weak anti-poaching laws and in hiring more game rangers, some of whom are ex-poachers who already knew tigers and tiger habitat better than anyone else.

The work of Hornocker, Quigley, and their fellow Russian biologists is proving positive for the survival of the Siberian tiger. The subspecies's population today is estimated at just 200 to 400 big cats, but there is hope that the population could recover and actually increase in the future.

Some tiger country, as here in Assam, India, is shared with equally endangered Indian elephants.

A hungry tiger crouches at water's edge to study the animals that have gathered here to drink.

All photos: *A tiger watches as a sambar stag wades neck deep to reach succulent vegetation. Suddenly, the crouched tiger explodes out of its cover and hits the water directly toward its target. Whether it scores depends on how quickly the deer can get out of the water and escape.*

Opposite page: *Of all the great cats, tigers share with jaguars the greatest affinity for water and wetlands. They are strong swimmers.*

Top left: *In India, the splendid peafowl is both a nemesis and an occasional meal for tigers. The birds scream warnings of a cat's approach and just as often are caught and eaten.*

Center left: *The abundant Hanuman langur monkey is a common prey species of the tiger.*

Left: *In Assam, India, the water buffalo is a major prey species of tigers. Poachers also use tethered buffalo calves to lure the cats into rifle range.*

Above: *The chital or spotted deer may be the most available and abundant prey of Indian tigers the year around.*

Opposite page: *An adult tiger pauses before crossing a pond in Kanha National Park, one of India's finest tiger reserves.*

Above: *Mating is often a lively, seemingly violent ritual among tigers. A moment after this picture was taken, the cats were calm and lolling side by side.*

Opposite page: *A baby tiger about five weeks old waits for its mother to return from a hunt.*

Above: *Nearly half grown, this tiger cub gambols around a shallow pond to develop hunting skills.*

Left: *About six weeks of age, this baby tiger risks a trip to a pond while its mother sleeps nearby.*

Opposite page: *Despite conservation programs and the establishment of reserves in Siberia, the tiger's habitat continues to shrink.*

Above: *A full grown, healthy Siberian male tiger such as this one is the world's largest cat.*

CHAPTER 5

SNOW LEOPARDS

The Ghost Cats

During the summer of 1976, when our legs and lungs were a lot younger, Peggy and I trekked across Kashmir in northernmost India. We walked the lofty, bleak, long-disputed Himalayan borderland shared with Pakistan and China, which was more peaceful then than now, although ethnic tensions were growing. Our goal was to photograph the awesome scenery and the wildlife, but we had little luck in spotting Kashmir's wild animals. We found a few colonies of wary marmots and scattered bands of bharals, or blue sheep, but they were never nearer than a mile away. The only evidence we found of the snow leopards living here on top of the world were faint tracks in the dust one morning near the 13,000-foot (3,900-m) Yemher Pass.

Weeks later back in Srinagar, Kashmir's capital, Peggy and I explored some of the markets that lie beside beautiful Dal Lake. Among the exotica we found prominently displayed for sale in several shops were snow leopard pelts. Selling, even possessing, one of these pelts was illegal, but shopkeepers assured me we would have no problem getting one back into the United States. We said no, thanks.

I still have never seen a snow leopard living wild in its steep, rugged range above treeline in central Asia. Because of its scarcity, reclusive lifestyle, and remote habitat, this cat rivals the jaguar as the most challenging of all large felines to locate, let alone study. Even if you do come near a cat, its long, thick, smoky-gray fur coat patterned with dark spots and rosettes makes it difficult to see when it is crouched motionless in its hazy gray or snowy surroundings. The snow leopard, *Panthera uncia* or *Uncia uncia,* is also commonly called the ounce, but its nickname may best describe it: ghost cat.

The existence of the snow leopard was unknown to the Western world until two centuries ago. In 1761, French naturalist Count Georges Buffon published a book with a color drawing of a spotted cat closely resembling the snow leopard, yet the accompanying text seemed to better describe the habitat and behavior of a cheetah. The London Zoo received an ounce in 1891, the first ever seen outside Asia, but the cat did not stimulate much interest. Not until 1903 did a live snow leopard reach the United States, a gift to the New York Zoological Park from a trustee. It soon escaped and was shot by a city policeman.

In his 1907 book, *The Game Animals of India, Burma, Malaya and Tibet,* author Richard Lydekker dismissed the snow leopard as a true ghost that didn't exist. Until fairly recent times—in fact until the same George Schaller, who had

A snow leopard studies its vast, lofty territory from a lookout on top of the world.

already studied lions, tigers, and jaguars, began his own search for the ounce in Pakistan in the 1970s—the snow leopard was a mystery.

Finding that first snow leopard was Schaller's greatest challenge, much like finding the proverbial needle in a haystack—except that this haystack was in near-freezing weather at an altitude of 3 miles (4.8 km) high. "I ventured into the mountains," Schaller wrote in his 1977 book, *Mountain Monarchs*, "with the hope of studying snow leopards, but my attempts failed, as almost perversely the animals eluded my efforts to observe them." The indomitable biologist did not give up. Eventually, he was able to follow one female for about twenty-eight hours over one week. And almost certainly he was the first to photograph a snow leopard in the wild. Schaller's work seemed to ignite an interest in the species that is still growing today. Our knowledge is also growing, albeit slowly.

The snow leopard is stocky and slightly smaller than a spotted leopard. Younger females weigh about 60 pounds (27 kg) whereas an older male may weigh as much as 120 pounds (54 kg). Snow leopards have short legs, broad paws, and tails as long as their bodies.

Snow leopard range is vast, but the boundaries are not certain. This range extends from the southern Himalayas and most of Tibet westward over northern Pakistan and into Afghanistan, northward through eastern Russia and into Sin-

kiang, western China. Another, probably separated population lives mostly in Mongolia. Usually found between 9,000 and 14,500 feet (2,700–4,350 m) elevation, snow leopards have been reported up to 18,400 feet (5,520 m) and as low as 2,800 feet (840 m) above sea level.

The best snow leopard range is just along the treeline or well above it, where the deepest gullies, steepest cliffs, and highest ridges discourage all but the most determined and motivated humans from following. The cats' thick winter coats and large paws covered with insulating hair make it possible for them to survive where winters are too brutal and long for a person to wander far with heavy capture equipment. They are most active around dusk and dawn when human hunters are wise to sit beside their own campfires. The population density of the cats seems low throughout their range, and they can only be regarded as endangered wherever they stalk in the thin atmosphere of high-altitude central Asia.

Much of the earliest interest in the snow leopard emanated from the old Soviet Union. During the early 1960s, Russian biologists became concerned about the decrease in cat numbers, as evidenced from declining sales of the furs. Limits on hunting and capture of the animals were initiated in Kazahkstan and other provinces, but were largely ignored. Wearing a leopard-skin cap and jacket was long a symbol of daring, bravery, and authority in mountainous rural Asia.

At times snow leopards venture downward to hunt in lower elevations, especially in Bhutan and Nepal.

In early reports and recommendations on snow leopards, some Soviet biologists commented on the cat's "playfulness and trusting nature." Villagers in remote areas, however, considered them savage and extremely dangerous. Still, there are only two references of snow leopard attacks on people anywhere, both in the Alma-Ata region of the former U.S.S.R. In 1940, a leopard attacked two people, was killed, and on later examination found to be rabid. Another cat leaped from a cliff onto a man passing below. It also was tracked down, killed, and found to be old and almost toothless. One summer in the 1970s, rumors circulated of a dangerous leopard lurking and one night attacking a Swedish hiker in Nepal. This animal simply disappeared and the incident was forgotten.

The Langu Gorge region in western Nepal contains some of the most remote, forbidding terrain in the Himalayas. As recently as 1964, only forty or so families lived there, and most had never seen a foreigner. In 1984, biologists Rodney Jackson and Gary Ahlborn decided the Langu Gorge was the perfect place to study snow leopards. During a ten-day walk over dangerous trails from the nearest crude airfield, they packed in enough food and scientific gear to set up camp for a major study of the cats. Then followed the most severe winter in the memory of the locals, and months passed before enough snow melted for the pair to begin work. Eventually, they were able to set out a leg snare made of aircraft cable baited with a live goat. Both men were resigned to a long wait.

Miraculously, the next morning the trap held a leopard. Twenty minutes after tranquilizing it, the world's first radio-collared snow leopard walked away. Six more months passed before the biologists captured the second of five they caught altogether. What they learned during four years of tracking these animals over the mountain terrain turned the danger and terrible hardships into high adventure and a hugely rewarding experience, both professionally and personally.

The radio-tracking telemetry quickly revealed the leopard's preference for steep, broken terrain, and especially for cliffs. Such high places were daytime bedding sites and vantage points from which the cats watched for prey, mostly the blue sheep that grazed across open terrain. Jackson believes that "crag leopard" would be a better name for the cat.

Surprisingly, an individual cat's home range was small within this vast region. One of the collared animals remained inside a 5-square-mile (13-sq-km) area. Another territory was just three times that size, although still small. Unlike the other great cats, large males did not drive rival males from their territories, which often overlapped. Not only were the territories surprisingly small, but each cat also spent more than half its time in only a quarter of its total range. The biologists were surprised to learn there were certain popular areas shared by many cats, but their visits were staggered enough to keep plenty of distance—a mile or more—between them. The men concluded that this avoided fighting and injury.

The avoidance was possible by leaving "calling cards" where they couldn't be missed. All visiting cats scuffed the ground with their large paws or left behind visible scrapes. They urine-sprayed rock faces to identify their sex, status, and reproductive condition to other cats. Local villagers were amused by the way the biologists meticulously recorded every scrape, marking each one with red nail polish and plastic-taped bamboo sticks. They laughed when the men sniffed boulders to detect the leopard urine. One conclusion drawn from all this scrape studying and sniffing was that leopard marking increased in winter and early spring, coinciding with the breeding season. It enabled mating pairs to meet at the same time it persuaded others to keep a distance.

One of their study conclusions was that, despite an individual leopard's small territory, setting aside parklands or reserves alone might not enable the big cats to survive in the wild. This cat at the top of the alpine food chain needs more living space and wild prey than a small sanctuary can provide. When wild prey disappears, snow leopards naturally turn to the domestic animals that now wander almost everywhere in their range.

Without doubt one statistic is more revealing than everything else in the Jackson-Ahlborn study. Although they were in radio contact almost daily with one or more of their five collared cats during their four years of work, they only saw the cats a total of eighteen times, or just one sighting every two and a half months.

One of those sightings, however, was out of the ordinary. One morning they heard a bharal whistle in alarm. A split second later it was lunging downhill at full speed with a snow leopard right behind. The sheep stopped suddenly and pivoted away, losing only some hair from its rump as the cat flew overhead past its quarry to land on the rocks far below. Watching such attacks is not unusual for lion researchers in Africa or for tiger observers in India, but high in the Himalayas it is a once in a lifetime experience.

Virtually everything we know about snow leopard mating behavior comes from observing captive animals. Female estrus lasts from two to eight days. One female at the New York Zoological Park permitted a male to mount her on three consecutive days. During the second day, the male mounted twenty-two times. Most of the copulations were initiated by the female, which encircled the tom and rubbed against him. Usually the male grasped the nape of the female's neck in its teeth. Both growled or meowed on reaching climax.

There are few records of humans, either modern researchers or native hunters, finding a natal den between April and June when two or three young cubs are born. Females line these dens with their own fur, but even with the added warmth, natural mortality is probably high. From captive animals we know that gestation is from 90 to 100 days, and that young leopards weigh about 1 pound (.45 kg) at birth. Leopards become sexually mature between two and three years of age. In captivity, they might live to twenty years or more. Longevity in the wild must be much shorter.

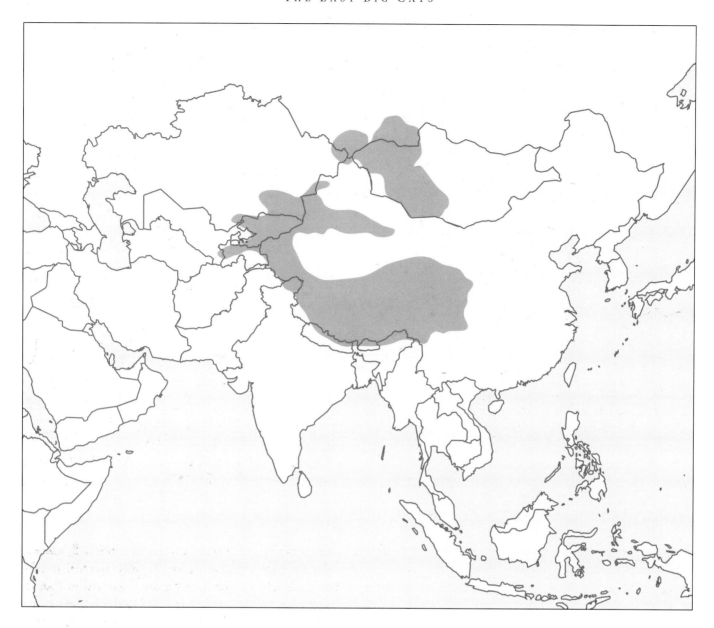

Above: *Range of the snow leopard,* Panthera uncia *or* Uncia uncia

Left: *Track of the snow leopard,* Panthera uncia *or* Uncia uncia
Front adult paw length: approximately 4 inches (10 cm)

Adult leopards in prime condition are able to kill both wild and domestic mammals up to three times their own weight. The natural prey species throughout their great range includes bharal, tahr, ibex, markhors (especially young ones), argalis, hares, musk deer, marmots (in summer only), and Himalayan snowcocks. At lower altitudes, they also prey on gazelles, wild boars, red pandas, tragopan, monal pheasants, chukkar partridges, and other birds. They will also scavenge any dead animals they happen to find. In Pakistan, Schaller noted forbs and grass in droppings. There are instances on record of snow leopards entering livestock sheds at night and slaughtering goats and sheep inside. It is not unusual for a cat to develop the habit of lurking around a village at night to snatch unwary watchdogs.

After making a kill or finding a carcass, a cat must closely guard the meat. Soon griffons, crows, magpies, or lammergeiers will arrive and make short work of unattended meat. In fact, biologists have located snow leopard kills by first spotting the circling birds. And there is one report of a cat running directly to a high ridge where birds were descending toward a yak that had died there.

In 1987, Nepalese zoologist Madan K. Oli began to study snow leopards by monitoring populations of prey species, especially of the blue sheep. Based in a lonely camp at 13,500 feet (4,050 m) in the rugged, upper Marsyangdi Valley of the Manang district, he also explored human impact on the cat. It is likely that the greatest and growing danger to snow leopard sur-

Blending well into its rocky background, a motionless snow leopard is extremely difficult to see.

vival everywhere is not from shooting and trapping them. Instead, it comes from loss of prey species that are unable to compete with increasing numbers of domestic livestock. Oli retrieved and examined carcasses of domestic animals killed by leopards, and interviewed the farmers of the region. What he learned was grim, considering that wherever in the alpine world the snow leopard survives today, the land is also occupied by a growing population of humans and livestock. Most people Oli talked with regarded snow leopards as competitors at best and vermin at worst.

A third of the households interviewed by Oli experienced losses to leopards during 1989 and 1990. These farmers lost about 3 percent of their livestock annually to leopards. Almost half of the cats' scats contained remains of domestic animals. Nearly all the natives interviewed expressed a dislike for the cats, and 87 percent favored solving their problem by eradicating all the leopards without delay. They sneered at the official "endangered species" status, and cat hunters or poachers were quietly honored community members. If snow leopards were unable to retreat into their remote and wild terrain, they might have been eliminated long ago.

Despite its innate ability to avoid humans most of the time, there are numerous reports of the cat's surprising tolerance of people. A hunter recalled shooting at one cat three times as it fed on a dead cow. The hunter continually missed, and the leopard kept coming back to finish its meal. A trekking guide told me that another snow leopard in Ladakh looked directly down on his group of twelve hikers passing in single file just 30 feet (9 m) below. One night, George Schaller crawled into his sleeping bag and fell asleep unknowingly about 180 feet (54 m) from where a female fed on her kill.

If snow leopards are doing poorly in the wild, the exact opposite is true in captivity. Thousands are living and breeding today in zoos, private collections, and game farms in Europe and North America. They have been proposed as fur farm animals and for reintroduction into the wild, neither of which is likely.

The fur farm idea may seem unacceptable, but it could be an alternative to the killing of wild cats to make coats. As recently as 1999, and despite protection by Nepal's own National Parks and Wildlife Conservation Act of 1973, snow leopard coats were still being sold in Kathmandu fur shops. Each coat required at least four cat pelts, and the average retail cost was $5,500. As long as such a market for snow leopard fur exists in the Western world, it is difficult to condemn a herder in fear of losing his livelihood in the Developing World.

For centuries, snow leopards have thrived on lonely landscapes such as this in Kashmir. But poaching and bloody conflict between India and Pakistan may have eliminated most or even all of the cats in these areas.

Smallest of the world's great cats, the ounce, or snow leopard, somehow survives in the most severe environment on earth.

Above: *An ounce races at full speed down a snowy slope toward a selected target.*

Opposite page: *A snow leopard drinks from a cold mountain brook, alert for both prey and enemies such as a herder and other human hunters.*

Right: *Markhors, especially young ones, are a food source in the highest and steepest mountain ranges.*

Below: *Snow leopards can survive in the highest elevations of the Himalayas wherever there is a large enough prey base of wild goats, wild sheep, and mountain marmots.*

Young, attractive leopards such as this one, captured alive, are popular in zoos around the world.

CHAPTER 6

COUGARS

The Cat of One Color

On a cool evening in May 1998, on the edge of Issaquah, Washington, just outside Seattle, a mountain lion materialized in the backyard of a suburban home to attack and kill a family's pet Labrador retriever while two children watched in horror from the safety of their doorway. The big cat then dragged the 50-pound (22.5-kg) dog into an adjacent woods.

If that wasn't enough to shake up the family, what happened early the next morning did.

State game warden Rocky Spencer and a professional hunter with hounds arrived to track down and kill the dangerous cat. They were not optimistic about finding the cougar because it had plenty of time to be miles away by then. But this cat had no intention of fleeing. The hunters found it only 100 feet (30 m) away from where it was last seen, and where it had partially eaten and buried the pet. Instead of running, as cougars normally do when faced by a pack of dogs, it attacked them savagely enough to scare off Spencer and the hunter and send the hounds scurrying back to their kennel truck to lick their wounds.

A second hunter with fresh hounds arrived. Surely this time, they believed, the cat would head for the mountains not far away. Instead, they tracked it just a short distance away and this time shot it. The cougar died in a spot where neighborhood children often play, and where the suburban sprawl of Seattle meets the forested foothills of the Cascade Mountains.

On examination, Spencer found that the cougar was not a sick or injured one unable to catch natural prey. It had not been intimidated by the presence of people, but rather had become habituated to living near them—the Labrador retriever was only lunch.

At almost the same time, another cougar was making news on Seattle television and in daily newspapers. This one found it easy to enter Northwest Trek, a popular 1,500-acre (600-hectare) wildlife park in nearby Eatonville. During a two-week period, the cat killed several deer and bighorn sheep inside an area enclosed by a 10-foot-high (3-m) fence until Spencer and another pack of hounds finally treed it. The 130-pound (58.5-kg) male cougar was also a healthy animal.

Once hunted nearly to extinction, cougar numbers are on the rebound in many sections of the United States and Canada. It is an ecological success story that many are celebrating in times when such successes are rare.

Yet the news about cougars has others watching nervously over their shoulders. There are worries that, unlike some of the other big cats, this one is getting far too comfortable with the booming human populations that everywhere are invading wildlife habitat. More people and more

At cool daybreak, a cougar appears at a mountain pond where deer and elk often come to drink.

cougars inevitably mean more encounters, some of them deadly. Of the eleven fatal cougar attacks recorded since 1890, half have occurred during the last decade. Nonfatal attacks also are on the rise, as are reports of the cats preying on pets and livestock.

The cougar or mountain lion, *Felis concolor*, the "cat of one color," has as many aliases as controversies following its trail. It is known by a variety of names over its vast terrain in North and South America: puma, panther, painter, catamount, *léon*, and cinnamon cat are just a few of them. By any name it is an adaptable animal, cunning predator, and superb athlete. The American wilderness would be less exciting and an incomplete place without it.

When Christopher Columbus first waded ashore in the New World, the cougar was the most widely distributed large carnivore living in the hemisphere. The cats ranged from what is now the Yukon Territory in the Canadian northwest to New Brunswick in the northeast and southward all the way to the tip of South America. Cougars thrived in every conceivable environment in between: mountains, deserts, coastal and subalpine forests, swamps, fertile river valleys, and on the Great Plains. Records show that cougars existed in all Canadian provinces and in each of the United States except Alaska and Hawaii. That original range is greatly reduced today and, with one exception, is limited to the western United States, Canada, and recently, southeastern Alaska.

A mother cougar bathes her kitten.

Thirty to fifty Florida panthers, *Felis concolor coryi*, survive in extreme southern Florida, thanks to the protection of the Everglades National Park, the Big Cypress National Preserve, and Fakahatchee Strand State Preserve, which together total about 500,000 acres (200,000 hectares). This Florida subspecies once ranged throughout the southeastern United States from the Atlantic Coast to eastern Texas and north to Tennessee and South Carolina. Yet this big cat was either ignored or regarded as vermin in a tropical ecosystem we were losing. The Florida panther was officially listed as an endangered species in 1967.

Data collected in 1989 from twenty-nine radio-collared Florida panthers indicated that the survivors were losing genetic diversity—similar to the African cheetahs—and that extinction by 2030 was likely. In 1992, with the health and number of Florida panthers still in decline, biologists made a controversial decision. Several female Texas cougars, their closest relatives, were trapped and introduced onto cypress hammocks in the southern Florida sanctuaries. Hybrid litters have since been produced, and some inbreeding problems might have been corrected. But too many of the cats are still killed by speeding cars on the busy trans-Everglades highways. Fatalities from territorial conflicts also have troubled researchers.

An increasing number of sighting reports of the eastern cougar, *Felis concolor cougar*, come every year from West Virginia to Nova Scotia. The eastern cougar was common two centuries ago, but is now usually considered extinct. Some, maybe most, of these recent sightings are cases of mistaken identity of coyotes, mongrel dogs, or even bobcats. Some also might be of escaped captive cougars, as indicated by tracks. But there is hope that somehow a few of this little-known subspecies may have survived in this region that now teems with white-tailed deer, always its natural prey. In fact, an explosion of white-tailed deer across America seems to be attracting cougars eastward. In 2001, a cougar was killed on an interstate highway in Kansas. And sightings have been confirmed in Michigan and Minnesota.

With the inexorable march of humanity westward across North America beginning in the eighteenth century, native wildlife was and is still the victim. The vast bison herds disappeared within just a few decades. The last survivor of millions of passenger pigeons died in the Cincinnati Zoo in 1919. The cougar also came close to being eliminated.

In 1960, only an estimated 2,500 to 3,000 cougars were believed alive in the remotest wilderness areas of the Rockies and Cascades. Commonly regarded as an enemy of humans and especially of our mushrooming livestock industry, cougars were not simply hunted without restriction, they were also trapped, tracked, poisoned, and shot from aircraft by government professional hunters, and bounties were paid for their killing. The goal was to wipe them out. Even America's pioneer conservationist, Teddy Roosevelt, despised the mountain lion. "The big horse-killing cat," he wrote, "the destroyer

A young cougar explores the pine forest along the northwest Montana-Alberta border.

of deer, the lord of stealthy murder, facing his doom with a heart both craven and cruel." That rhetoric was composed in the early 1900s, but was quoted as recently as 1969 during a stockman's meeting in Montana to request more federal aid to kill the cats.

Of course, cougars do occasionally prey on livestock, and these individuals can be eliminated. But cougars have never been a significant threat to people. Compared to other dangers facing modern humans—alcohol abuse, drunken drivers, random violence, even sunburn, obesity, bee stings, lightning strikes, and domestic dog bites—cougars are practically a benign presence. The good news is that national attitudes have radically changed in the last decades of the twentieth century. Most Americans and Canadians now see the cougar at least as a vital and attractive survivor, a symbol of wild places and essential to the balance of nature.

In the scattered archives of the first settlers, cougars al-

ways were recorded as shy animals that avoided people. But by the time bounty hunting ended in the 1900s, the history of persecution left us an extraordinarily furtive creature that was more a mystery than a menace. A keen outdoorsperson—rancher, hunter, or backcountry hiker—can spend a lifetime in the best mountain lion country and never see one or even any sign of one. For many years, Peggy and I lived on the edge of Montana's Absaroka-Beartooth Wilderness Area, which, with a good population of deer and elk, is excellent cougar country. Every winter, a cougar left tracks in the snow while encircling our house at night. One January, a cat killed and ate a white-tailed doe within sight of our house and another about 200 yards (180 m) beyond the back door. But we never once had a glimpse of this ghostly cat. It is thus no wonder that until recently little was known about this elusive wild cat.

* * *

Above: *Track of the cougar,* Felis concoloro
Front adult paw length: approximately 4 inches (10 cm)

Right: *Range of the cougar,* Felis concolor

In 1964, biologist Maurice Hornocker began the first of his important studies of several big cats. For five years, he researched cougars in the most exhaustive and revealing studies of the species yet attempted. It was also a test of his own physical stamina and determination, as well as a dose of high adventure.

Hornocker had gained valuable experience dealing with large mammals as an assistant to John and Frank Craighead during their pioneering investigations of grizzly bears in Yellowstone National Park. For the cougar work, he selected the Big Creek drainage, an area of about 200 square miles (520 sq km) in the Salmon River Mountains of central Idaho. He enlisted three invaluable assistants: experienced lion hunter and rugged woodsman Wilbur Wyles and Wyles's pair of redbone hounds, Ranger and Red.

The two men and the dogs trekked after cougars for more than 5,000 miles (8,000 km) on foot and often on snowshoes in winter; it was the equivalent of walking a quarter of the way around the world. Together, they captured alive forty-six different cats. Each cat was tracked and treed by dogs in extremely steep and rugged terrain. They were then tranquilized, weighed, examined carefully, tattooed, tagged and/or fitted with a radio collar, and released unharmed. All this had to be accomplished with the minimal equipment—other than

food and wilderness survival gear—that the two men could carry on their backs in an area where the only paths are thin game trails, often obscured by snow in winter. But the information they obtained was fresh and invaluable, laying new roads to understanding cougars.

Like big cats everywhere, the resident adult cougars here had firmly established territories. The boundaries were marked with mounds of brush or pine needles scraped together and urine-scented during regular patrols. Females marked winter home ranges of 5 to 25 square miles (13–65 sq km). Males wandered over generally larger territories of 15 to 30 square miles (39–78 sq km). The Idaho cats seemed to avoid invading each other's territories, behavior called mutual avoidance, which is also a survival mechanism for such solitary creatures as these. Prides of African lions share food killed by others of the group, but each mountain lion is its own sole means of survival. In fact, the mountain lion is the most solitary and self-sufficient of all the world's big cats.

The average female cougar in Hornocker's study weighed 100 pounds (45 kg) while the average male weighed 150 pounds (67.5 kg). The largest tom he weighed was 181 pounds (81.45 kg) and the lightest 130 pounds (58.5 kg). Mountain lions might weigh slightly more elsewhere on the continent, but the rumors of 300-pounders (135-kg) are just tall tales.

The largest ever recorded anywhere on honest scales weighed 276 pounds (124.2 kg).

The Big Creek study area accommodated a stable resident population of only ten lions, or one per 20 square miles (52 sq km), and this just might be a maximum anywhere for the species. At times, transient cougars wandered across boundaries and across the study area, but they never lingered. Each year during the study, two or three litters of two or three cubs were born. But still the population of the area remained at about ten cats.

Males and females seem programmed to spend as little time together as possible during courtship and mating, which can occur at any time of year. A female coming into estrus is not secretive about it, frequently caterwauling—shrilly screaming—in the night. Similar to but more penetrating than the cry of housecats mating in backyards, a caterwauling cougar makes one of the most eerie sounds a wilderness camper is likely—or lucky enough—to ever hear.

To date, no wild mother cougar has ever given anyone the chance to see, let along study, how cubs are raised. The little we know about this behavior comes from captive animals in breeding compounds. Maybe because they are not free to do much else, captive mothers spend much time grooming themselves and playing with their young. If too closely watched, they often pick up the cubs by the scruff of the neck and carry them elsewhere, a trait shared with many other carnivores.

After a gestation period of ninety days, as many as six cubs, but usually two or three, are born in a cave or secluded den. The young, unlike their parents, have distinct dark spots on their soft brown bodies that disappear as they mature. Competition among cubs begins soon and the survival rate is not high. Cubs remain with their mother for almost two years of training. Fathers never participate in feeding, protecting, or training their own offspring. Young cougars are self-sufficient by eighteen to twenty months, and then strike out to find territories and a life of their own.

The main prey of this brown cat during Hornocker's study was mule deer as well as some elk. But that has gradually been changing as white-tailed deer take over mule deer range in the West. Hornocker and Wyles examined scores of cougar kills in five years, and it was evident that young and old animals, the easiest prey, composed most of the total kill. Specifically, three of every four elk and almost two of every three deer were less than one and one-half or more than nine and one-half years old. The Idaho lions also commonly ate anything else they could catch: hares, squirrels, raccoons, coyotes, even mice and grasshoppers. Quills found imbedded in feet and muzzles were evidence that they try to take porcupines, too. They also unfortunately but naturally hunt endangered species.

Directly north of the Big Creek study area, on the border between British Columbia, Idaho, and Washington, is an endangered herd of mountain caribou whose numbers in the late 1990s seemed to go into free fall. The culprit, at least at first, seemed to be cougars. In one parcel of land near Kootenay Pass, most of the deaths of twenty-four collared caribou that occured within a short time could be traced to cats, and there was international pressure to "do something." But what seemed obvious has become a baffling mystery. While local hunters claimed the region was swarming with cougars, researchers have not been able to locate many at all. It's a scenario that has happened elsewhere, including on Vancouver Island, British Columbia.

Mountain lions seem to hunt as much or more in daylight than in darkness. A healthy adult is a splendid hunter, agile, capable of sudden bursts of speed, an excellent climber in trees as well as steep rock cliffs, a swimmer when necessary. Veteran cougar hunters have claimed the cats can leap 20 feet (6 m) into trees and drop back to the ground from that height without injury. They can leap horizontally as far as 45 feet (13.5 m).

All of a cougar's senses are keen, even compared to the other big cats. It is evident when watching a captive animal that it is well aware of activity all around without turning its head. Its vision is stereoscopic, which means it can see sharp, three-dimensional images, ideal for hunting in sunlight as well as after dark. The cougar's highly developed senses, especially its hearing and vision, explain why it is so seldom seen by humans. A cougar is as an efficient stalker and killer as any of the other big cats. But driven by hunger during the hard times of winter, a cougar might take on more than it can handle.

One day in 1967, Wyles came upon a female that had been injured when attacking a small bull in a herd of elk. From evidence in the snow, Wyles could read how the two had rolled down a mountainside together, finally crashing into a tree, which permitted the prey to escape. Wyles treed the cat, which, although bloody around the face, appeared otherwise unharmed. Three weeks later, Wyles recaptured that same female about 4 miles (6 km) away. Now, she was thin and barely able to climb out of reach of the hounds. A quick check revealed a broken jaw, canine teeth torn out of the jaw, and antler punctures in the shoulder and hind legs. The cat was clearly starving and suffering horribly, and Wyles mercifully shot her.

There is one characteristic weakness that has often been the undoing of this formidable cat: Rarely will a cougar face up to a barking dog, not even a yipping lapdog that it could quickly destroy, let alone a whole pack howling in hot pursuit. This has been a boon to poachers even more than to biologists. Most of the other big cats will normally turn on and sometimes kill pursuing dogs; leopards are especially likely to retaliate. But the average cougar will come to bay in a tree rather than fight. Large males as well as females with cubs have been treed by terriers that weigh little more than jack rabbits. High in a tree or canyon wall is safety enough from dogs, but is no escape from a hunter with a rifle.

More often, however, things go right for cougars. One winter afternoon in 1992, photographers Jim Mepham and Steve Torna were focusing on mountain goats at a mineral lick in Glacier National Park, Montana. Suddenly the goats bolted, leaving one behind in the grasp of a mountain lion, which killed it. As the cameramen watched, the lion pounced on a second nanny and both went rolling far down the slope. At the bottom the cat was on top of its dying prey. It remained in the area for four days until most of the goat meat was eaten.

Quite often cougars are blamed when animal populations, especially mule deer numbers, begin to fall in a given area. Without doubt too many cats can cause a deer decline, especially in places where too many deer are living on already poor range that has been overgrazed by livestock. But Hornocker observed something else in lion-deer relationships: Whenever a lion kills a deer, the herd often moves far away and remains there until it is again molested and forced to relocate. In this way the deer (or even elk) do not overbrowse certain areas as they might otherwise, causing serious damage to their own food supply. The result is healthier, as well as more wary, deer.

California has long been a battleground over the status of the mountain lion. Public opinion sways wildly in one direction, then the other. There are plenty of supporters here for ridding the state of cats entirely and an equal number who would protect all cougars absolutely. The state's bounty system was not ended until 1963. Only six years later, in 1969, the cat was classified as a regulated big-game species. Then two years after that, a moratorium on killing cougars, except for individuals that were caught eating livestock, was voted into law. This ban has been blamed for dwindling game herds from Mexico to Oregon. California's North Kings deer herd did drop from an estimated 17,000 animals to less than 3,000 two decades later. Encroachment of new suburbs and poaching also was responsible for the decline in the herd's numbers, however.

An accurate count of cougars in a state as large and diversified as California is virtually impossible. And the cats keep making headlines and hardening public opinion by appearing regularly on the outskirts of San Diego—scattering joggers—and even in the state capital of Sacramento, at exactly the same time that state lawmakers were debating what to do about them.

Scientific evidence is mounting that lions are a grave threat to the endangered bighorn sheep that cling to existence in the sheep's scenic winter range near Bishop, California. Biologist John Wehausen has studied these sheep and mountain wildlife during most of his long career and verifies this. During a field trip with him, we saw few sheep, but he pointed out places in brush drainages where he had found cat-killed carcasses, and he has witnessed several actual kills.

After a short chase in powder snow, a cougar catches a snowshoe hare.

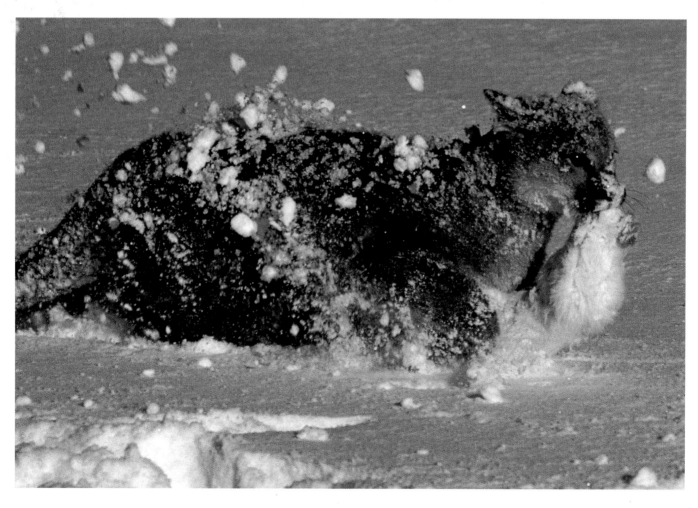

Despite the fact that biologist teams in almost every state from New Mexico to Alberta are now following lions with the latest electronic equipment, conflicting information about their target animal is being received. Most males do keep strictly to established territories, but cougar society has its share of vagabonds. Sheepherders in Sinaloa, Mexico, shot one that had been tagged far away in northern New Mexico. Wildlife biologist Kenneth Logan marked a two-month-old cub in Wyoming. It was killed two and a half years later near Aurora, Colorado, about 300 miles (480 km) away as the crow flies. If nothing else, this wide wandering will assure genetic diversity for the cat.

The world knows next to nothing about cougars living in Central and South America where, as elsewhere, ranchers hate them and train packs of dogs to keep their numbers low. South American cougar pelts tend to vary more in color than North American cats. They can vary from region to region, from reddish to dark brown, tan to almost gray. Age may also make a difference, with the pelts of older cats averaging slightly paler than young ones.

Although the lions still probably exist in all countries from Mexico southward, we have little idea of their abundance or exact range. They are found in a great variety of habitats from moist tropical forests and *llanos,* or open plains, to above treeline in the Andes. Cougar tracks have even been recorded above 17,000 feet (5,100 m) in Peru. Those of Brazil's Pantanal share that alternately wet and dry environment with jaguars, but tend to avoid them as they do everywhere else. A problem wherever the two cats coexist is that it can be difficult to tell cougar scat and tracks from jaguar sign.

Chile is one bright spot for cougar study in the Southern Hemisphere. Hunting day or night, one cat that was tracked near Osorno killed fifteen pudus, a small native deer, in a 249-day period. During a longer Chilean study in Torres del Paine National Park, 409 scats were examined to learn about food habits. Mammals, largely European hares and mostly yearling and juvenile guanacos, appeared in 92 percent of the scats. One of the most spectacular places on earth, Torres is also anyone's best bet in South America to see a wild cougar.

In Panama, cougars are known to prey on collared peccaries, deer, pacas, agoutis, and snakes. In southwestern Brazil, they have consumed rheas (large flightless birds), capybaras, and tamanduas (lesser anteaters) as well as deer. Puma scats from San Juan Province, Argentina, contained eggshells, insect larvae, and beetles as well as guanaco and vicuña.

In many remote places, the lion is still feared as an angel of death. People tell in hushed tones that the cougar is able to hypnotize and paralyze prey, that it kills just for the fun of it,

Affectionate and protective of small kittens as are most felines, cougars will carry them to a safer alternate site if it seems necessary.

and that the big cat should be avoided at all cost. Many people still believe that the caterwauling of a female on a rainy night foretells a death in the family or hard times ahead. In Suriname, a fishing guide once gravely warned me that just to see a cat was a curse and that I should never mention it to anyone.

I have been lucky enough to see several cougars in the wild, the last one while hunting elk in Wyoming several years ago. I will never forget that brief glimpse of the skulking animal, golden in the late-September sun. Describing the moment to anyone who would listen has always been a pleasure.

Both photos: *Cougars blend well into the Utah red rock country near Zion National Park.*

Both photos: *The agile cougar is a powerful jumper—which is essential when traveling and hunting in the desert.*

Above, both photos: *Confrontation and snarling are part of the active courtship that precedes eventual copulation.*

Opposite page: *Natal dens are sheltered, secluded places, often in rocky slopes or in dark caves, as here.*

A mother and cub nuzzle each other.

Top: *How much time a cougar mother spends with kittens depends on how much time must be spent hunting for all to survive.*

Above: *Barely a month old, two baby cougars venture tentatively from their den.*

All photos: *An adult mountain lion moves silently and stealthily over the ground to approach its prey. Once it launches an attack on a deer or elk, a cougar becomes a tawny blur in the forest. Of all the great cats, only the cheetah may be faster than the cougar over a short distance.*

Above: *Once cougars roamed in every one of the United States, but they no longer hunt in the East or northern Midwest forests.*

Opposite page: *An Arizona cougar pauses briefly before dissolving into deep canyon shadows.*

Opposite page: *Gravely endangered, probably fewer than a hundred Florida panthers,* Felis concolor coryi, *survive in the Sunshine State.*

Above: *The scientific name* Felis concolor, *or "cat of one color," well describes the American mountain lion.*

Right: *The smaller bobcat,* Lynx rufus, *shares cougar country and is often mistaken for a young cougar.*

Below: *A white-tailed deer runs from a hunting cougar in the Rio Grande Valley of Texas.*

Opposite page: *Cougar country is colorful in autumn when the cats lurk along familiar deer and elk trails. Wherever they live—in mountains, forests, foothills, or desert edges—cougars blend well into their environment.*

Above: *Today, the cougar still ranges from the Yukon Territory of Canada, southward to Torres del Paine in southernmost Chile, where they prey on guanacos.*

Opposite page: *Though normally quiet, cougars are far more often heard than seen, usually during the mating period when making a sound that is eerily like a woman in great pain.*

CHAPTER 7

JAGUARS

The Cat That Kills With One Leap

On March 7, 1996, Arizona rancher and guide Warner Glenn and his daughter Kelly began the fourth day of a mountain lion hunt neither of them would ever forget. Early in the morning, one of their pack of experienced hounds jumped a "big lion," and a wild chase began across their remote Malpais Ranch. What followed was much more grueling than any lion hunt they ever recalled. The chase led directly up into the Peloncillo Mountains, steep, rugged, and brushy country where the boundaries of Arizona and New Mexico hit the Mexican border. This cat simply would not quit running and come to bay. The constant wind made it difficult for the dogs to follow the track and for Glenn, riding a tough old mule named Snowy River, to keep up with his dogs. Finally, after many miles, the cat found refuge atop a sheer bluff where the howling dogs below could not quite reach it.

When he arrived on the scene, Glenn could not believe his eyes. The animal snarling down at him from atop the bluff was not a mountain lion, but, incredibly, a jaguar. Fortunately, he always carried a camera on his saddlehorn and, while trying to restrain his hounds with one arm, snapped several photos of the jaguar with the other hand. Then just as sud-

denly as it began, the hunt ended when the quarry jumped down, slashed and bloodied the nearest dogs, and ran toward Mexico.

Altogether, it was a remarkable incident. Glenn had hunted lions and other game in that same borderland region for most of his sixty years without ever seeing a jaguar. Nor had his father ever found one in the twenty years before that. Before Warner Glenn's 1996 hunt, the last known jaguar in the United States was killed in southern Arizona's Dos Cabezas Mountains in the 1980s. In fact, only about seventy jaguars have been killed north of the Mexican border during the twentieth century, and most of these were before 1950. The jaguar Glenn sighted may still be roaming free in the American Southwest, and the handsome cats might now be returning to the U.S. range they lived in long ago.

The jaguar, *Panthera onca*, is the third largest in the family of felines and is native only to the New World, where it is the largest cat. The word "jaguar" may have come from Brazil where Guarani Indians of the Amazon Basin called it *yaguara*—an animal that can kill with one leap. Everywhere south of the United States today, it is known respectfully as *el tigre*. By any name, it is a powerful predator that averages larger in size and heavier in weight the farther south it is found.

It is a hot midday in central Brazil, and a jaguar dozes out of reach of most annoying insects.

The jaguar's range in 2000 extended from northern Argentina to northern Mexico. Once, this was a wide and continuous range, but it has shrunken greatly as human populations expanded. Today, the species is considered threatened, although jaguar numbers seem to be holding steady in scattered areas that are still unattractive to human settlement. In fact, the worldwide conservation community, some governments, and natural history scientists have become increasingly interested in saving jaguars. And none too soon. As with the tiger in India, the jaguar in the Americas is an ideal symbol that creates interest and generates funding to keep intact natural areas and to maintain their biological diversity.

Jaguars have intrigued and instilled fears in people ever since the two have had to coexist. The cats were important in the art, architecture, cultures, and rituals of Central Americans from the Olmecs in about 1200 B.C. to the Mayas and Aztecs much later. Archaeologists are still unearthing temples and tombs devoted to jaguars. Modern museums contain evidence of jaguar societies, costumes, jaguar soldiers, and skeletons of humans sacrificed to feed the cat "gods."

A healthy, full-grown Brazilian tom might measure more than 8 feet (240 cm) from its nose to the tip of its tail and approach 300 pounds (135 kg). While most other large cats often kill prey with strangling throat holds, jaguars can kill some prey animals with one bite through the skull or neck.

Superficially, this largest of New World cats resembles the leopard of the Old World. Its yellow-to-orange fur is decorated with similar black rosettes, but a distinguishing feature of every jaguar is the black spot in the center of each rosette. As in leopards, all-black or dark-brown color phases occur occasionally among jaguars. In bright sunlight the pattern of rosettes can be detected even on the dark melanistic cats.

Although jaguars rarely if ever hunt in trees, they are powerful climbers, which can be helpful when tracked by a poacher's dog packs.

Jaguars have shorter legs, a stockier stance, more muscular chests, and appear heavier in the body than the lithe leopards. Jaguars are probably also slower afoot, but are stronger swimmers and therefore better adapted to living in green wetland basins. They also inhabit the edges of deserts, *llanos* or open plains, dry woodlands, rain forests, and—too often for their own good—ranchlands and the fringes of cultivated areas.

Currently, jaguars are the least studied and therefore the least understood of all the big cats. Biologists generally agree that adults are solitary and require enough food that they must wander over large areas to find it. Throughout their range, jaguars are known to prey on more than eighty-five different species, including capybaras and coatimundis, deer, armadillos, tapirs, iguanas, peccaries, fish, turtles and turtle eggs, snakes, birds, monkeys, frogs, and caiman. Their powerful jaws enable them to crack the shells of river turtles and land tortoises. No matter where a hungry cat roams—over the Pantanal grassland of Brazil, the dry thickets of Bolivia's Chaco country, or the steep canyons of Mexico's Sonoran Desert—any living creature is fair game.

Judging by how seldom they are seen during daylight, jaguars must be among the most nocturnal of all felines. Mating seems to take place at any time of year, and these are the only periods when adults might be found together. From observations of captives, we know that males and females in estrus spend a few days together alternately grooming, growling, and mating. Two to four cubs are born, and as with other cats, most do not survive. Professional hunters have found only a few caves and other well-hidden places where young have been born. One litter was found by a local fisherman on a small island in a Colombian swamp that the mother could cross only by swimming.

Jaguars are more likely to be heard than seen, and even hearing one requires spending long, uncomfortable hours at night in known cat country. Probably to mark territories or announce their presence, males primarily roar at intervals as they travel. Depending on distance and humidity, a roar can sound muffled or penetrating. Some indigenous woodsmen have learned to imitate the sound by blowing into an earthen jug or hollow gourd.

Once on a still, hot night beside the Cienega de Zapatosa, deep in Colombia's wilderness, fishing and hunting guide Macedonio Polo demonstrated to me how it is possible to "talk" to a wild jaguar. Grunting into his jug, he was able to call a jaguar closer and closer to where we waited. Then, suddenly, the cat must have detected a false note, and the rest of the night was silent. It was eerie not to know exactly where the cat was. Had it left, or did it lurk in the vegetation nearby?

An accurate count of jaguar numbers today is still lacking, and knowledge of their general behavior is superficial. While a fairly accurate census might be possible of an isolated region—say of a certain limited sanctuary or reserve—there is no way so far to count the cats across their entire range. A few thousand might be a good guess, but no more. Nor do we yet know much about the size of individual jaguar territories.

Roaming the Western Hemisphere in its present form for more than 50,000 years, the jaguar came near to being wiped out forever in the twentieth century. At the dawn of the 1800s, an estimated 4,000 cats a year were being taken by hunters in Spanish-colonized America. At first, killing jaguars was just a part of clearing the land of predators as well as of trees to raise cattle. But the settlers quickly learned that the dead cats were a valuable commodity. Most of the pelts were shipped to Europe where jaguar coats became the rage among the chic upper classes.

In the two decades after World War II, 13,000 of the luxurious skins were exported from Peru alone. Brazil was exporting more than 6,000 pelts annually. Even as recently as the 1960s, about 15,000 pelts were still reaching commercial markets overseas. Between 1968 and 1970, the United States alone imported 31,104 hides. Environmental scientist Norman Myers estimated that the international trade in spotted-cat skins—ocelots and margays as well as jaguars—totaled at least $30 million a year. No wonder jaguars teetered on the edge of extinction before an international agreement to ban trade in most wild-cat skins finally took effect in 1975.

It is unrealistic to believe that the trade stopped entirely, even though the pelts of jaguars became more difficult to obtain. Illegal killing continues because of conflicts with livestock. But encroaching civilization and the increasing loss of habitat has been an even greater worry than the lucrative market for pelts. There was a good chance that we might lose this splendid species before we knew anything at all about it.

George Schaller, a pioneer in the study of African lions and Asian tigers, was also the first in 1977 to investigate jaguars in Brazil. But his work was not applauded or aided by ranchers in his study area as they did not share his interest in predators. In 1981, Schaller sent Alan Rabinowitz, who had just received his Ph.D. in wildlife ecology, to Belize to make a preliminary study there of the big cats. For two months, Rabinowitz roamed jungles, swamps, and dry brushlands, recording footprints and scrapings, scats and kills, and talking to back-country natives. He soon discovered one thriving population of jaguars and continued his investigations there.

For the next two years, the young scientist lived in an abandoned logging camp among a few Mayan families in the tropical Cockscomb Basin. Constantly in the field, exploring, and only rarely catching a brief glimpse of his elusive quarry, he studied the diets, habits, and territorial needs of jaguars. He also began to lobby Belize government officials to protect the unique and beautiful area. Thanks in part to his efforts, in 1984, Cockscomb Basin was designated the world's first jaguar preserve, a total of 102,000 acres (40,800 hectares) in the largest contiguous block of forest surviving in Central America.

Jaguars like to spend middays where they are least annoyed by insects, above the ground where air can circulate.

Rabinowitz observed that Cockscomb female jaguars ranged over areas of 2 to 12 square miles (5–31 sq km); male territories averaged twice as large. He also detected a decided cat preference here for thick natural forest over more open areas. The jaguars seemed to avoid human structures.

One of Rabinowitz's most surprising discoveries came from skeletal evidence taken from cats destroyed for killing cattle. Many of the skeletons revealed old gunshot wounds, But nonproblem animals that were live-captured had no such scars. The reason seemed to be that a bounty offered for jaguars then in effect encouraged forest workers to carry firearms, usually old shotguns. As a result, they often injured or crippled cats without killing them. Unable to catch wild prey, the wounded jaguars turned to stalking livestock. Information such as this obtained by Rabinowitz greatly changed Belize attitudes about the cat, which has since become a magnet for ecotours and tourists there.

In a vast region of Campeche, Mexico, National University of Mexico ecologist Gerardo Ceballos and Carlos Materola of Mexico's United for Conservation groups are also working to collect data on jaguars. Their studies are focused in the 1,787,000-acre (714,800-hectare) Calakmul Biosphere Reserve, established in 1989. Remote, hot, and considered too dismal for human occupation, it is the largest surviving rain forest in the Americas besides the Amazon Basin. Ceballos and Materola believe that no other reserve anywhere contains so many jaguars living as they did before Europeans invaded the New World.

Virtually the only human activity now authorized on the Calakmul Reserve is wildlife research with an emphasis on the top of the food chain, the jaguars. Biologists have been live-trapping, radio-collaring, and then tracking as many animals as possible for baseline data. Fortunately, they have

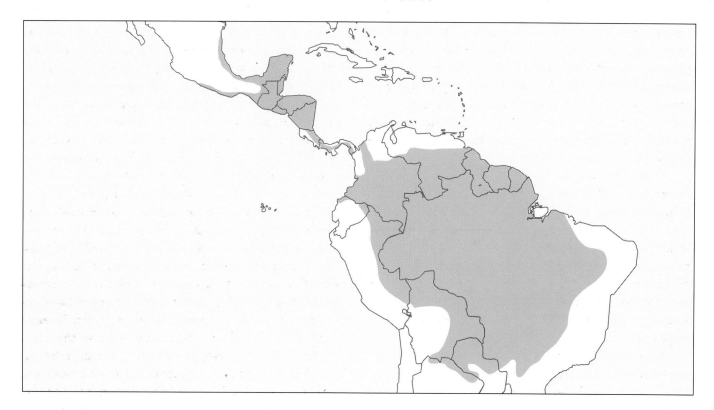

Above: *Range of the jaguar,* Panthera onca

Left: *Track of the jaguar,* Panthera onca
Front adult paw length: approximately 4½ inches (11.25 cm)

electronic equipment available that has already proved effective on the other great cats. Yet the reserve is so dense, so lacking in trails and roads, that to use the new gear, the biologists have to depend on experienced local woodsmen who once survived as poachers and hunting guides. The combination of modern technology and primitive skill has so far been working wonders. If jaguars have one best chance of longtime survival on Earth, many believe Calakmul just may be the place.

Before a radio collar can be slipped around a jaguar's neck, the animal must first be secured intact and alive. This is where those invaluable reformed poachers show their worth, employing ancient knowledge, feral instinct, and simple tricks of the woods. Calling the cats from tree blinds then enticing them into live traps with live animal baits is among the simplest means to catch jaguars in a region where a machete and a strong arm is necessary to travel any distance. Some older hunters are able to detect the passing of cats, or areas used frequently by them, by actually smelling the urine sprayed on faint trails and tree trunks. During a trip to the interior of Colombia in 1963, I met a blind old man, said to be in his eighties, who younger poachers led on a leash into the jungle where he was able to sniff out areas of jaguar activity.

Also on that trip, my guide Macedonio Polo described another strange but effective method they used to kill jaguars. A number of macaws were trapped alive and staked out around a jungle tree that was large enough to hold a man in a blind just above. Another macaw was tied by the foot and held in the tree blind with the hunter. Days might pass before a cat found the constantly squawking bait birds, but when it did, the tied bird was dangled down to the jaguar's face. When the cat turned its head up toward the fluttering bird, the hunter thrust his spear right into the cat's open mouth. This primitive trick has been used recently and effectively by biologists who substitute a dart gun for the spear and shoot the cat in the rump to tranquilize it.

Another important area attracting the attention of scientists and conservationists alike is the Pantanal of Brazil. It is probably the best place to see a jaguar in the wild. Here, in the states of Mato Grosso and Mato Grosso do Sul, is a basin fourteen times the size of the Florida Everglades. No place in the world can match the biomass of land birds—egrets, spoonbills, ibis, and storks as tall as the marsh deer wading among them. The Pantanal is also rich in mammals, reptiles, fish, and just possibly contains the most jaguars per square mile to be found anywhere.

During winter, the Pantanal is dry, a vast, grassy plain with gallery forests bordering the many tributaries of the Paraguay River. During normal summers, it is rainy enough to overflow most of the river banks, forming shallow lakes and marshes that are alive with creatures, but support few people. Because much of the Pantanal is flooded annually, only limited ranching is possible. In 2000, the seemingly conflicting interests of ranching and wildlife appeared to be in some balance. Hunting for the spotted cats was much easier in the Pantanal than in places as dense as Calakmul's jungle, but jaguar numbers have rebounded here greatly since the 1970s when unregulated hunting for hides was stopped.

One of the interesting new facts now known about Pantanal cats is that males especially have learned to use the all-weather raised roads connecting widely separated cattle ranches to more easily explore their territories of 10 to 60 square miles (26–156 sq km). Females with cubs range over much smaller territories. When the wet summer ends and the water evaporates or drains back into rivers, many Pantanal jaguars live on the bounty of fish stranded in shallow sloughs. Smaller caimans are also readily eaten at this time, as are many capybaras.

Sandra Cavalcanti of Utah State University is one of a new breed of energetic, enthusiastic wildlife researchers now busy in the Pantanal. Her focus is the study of the jaguar in relation to the livestock industry, as conflict still persists between the two. Alone, unarmed, and on horseback, Cavalcanti rides across the lonely land with a notebook, a water jug, and binoculars. In one case, she located a cow reported by a cowboy as a jaguar kill. Closer inspection revealed the truth: The cow died in calf-birth, and the nearest jaguar tracks showed a passing cat did not even pause for an inspection. When another rancher reported losing seventy cattle to cats, Cavalcanti investigated and concluded that only nineteen were jaguar kills. More—twenty-four—died of snakebites. Even when she finds a true cat kill, there is the problem of determining whether it was by jaguar or cougar, which also live throughout the Pantanal. Both cats reach their maximum size in this, the greatest of the world's last wetlands.

From the Pantanal northward, South and Central American countries have established about 200 nature reserves where jaguars still live. Almost all the reserves are too small or vulnerable to development, deforestation, and poaching to survive as intact ecosystems for long, but there is hope if many could be connected by natural corridors. To accomplish this, biologists have envisioned a wilderness greenway termed the *Paseo Pantera*, or Panthers' Pass. Just as saving the big cats elsewhere serves entire ecosystems and other endangered creatures, a *Paseo Pantera* through the Americas would also aid countless other, smaller creatures, including the threatened hyacinth macaw, Morelet's crocodile, dwarf porcupine, and a dozen different frog species. New computerized mapping techniques suggest that it is possible, and international wildlife conservation groups have helped fund work toward this end. But the downside is that the well-financed enemy—mining companies, lumbering interests, and the politicians who work with them—stand firmly in the way.

Consider the Calakmul Biosphere Reserve alone. It is the home of about 300 bird species, many rare and endangered. Its incredible richness is habitat for 550 species of vertebrates, 92 of them mammals, and about 40 species of bats. By 2001, ten jaguars have been captured or recaptured and fitted with radio collars here. These jaguars are tracked by jeep patrols, tall radio towers, or aerial surveys, and much solid information is collected and analyzed for future management needs. Biologist Gerardo Ceballos estimates that more than 240 jaguars now live on Calakmul, and that with continued protection that number might almost double.

In fact, surplus animals might even wander out of the reserve and repopulate other suitable habitat farther away. There are scattered instances of jaguars traveling vast distances, perhaps after being driven away in territorial disputes with other cats. One male was killed in 1955 near the southern tip of the San Pedro Martir mountain range of Baja California. It must have wandered across the entire Sonoran Desert, crossed the Colorado River, and then traveled south for about 100 miles (160 km)—a trip of 500 miles (800 km) from the nearest occupied jaguar range. There have also been reports of jaguars suddenly, briefly, roaming over Mexican ocean beaches on spring nights—once even on the illuminated fringes of busy Acapulco—and digging up the eggs of nesting sea turtles, always to vanish as mysteriously as they arrive.

Early in 1999, the Wildlife Conservation Society in New York convened a workshop of thirty of the world's jaguar researchers to determine what was necessary to protect the species through the next millennium. The group recommended one particular area, the 2.8-million-acre (1.12-million-hectare) Mamirauá Reserve in the Brazilian Amazonia, as another important locale to study. Located just beneath the equator where the Rio Japura joins the Amazon, the reserve is a huge area that, at the peak of the rainy season, is covered with water 30 to 40 feet (9–12 m) deep. There are a few human inhabitants who have learned to live here with that great natural fluctuation, and the Mamirauá Reserve allows them to use and market the ecosystem's natural resources in a sustainable way. They may fish for several edible species and hunt the forests for some birds and monkeys.

The Mamirauá is a paradise for many creatures. Caimans, turtles, manatees, and endangered pink river dolphins abound. So do the canopy dwellers, such as monkeys and sloths. But deer, armadillos, tapirs, agoutis, and other mammals must retreat to dry ground when the rains fall. Only one terrestrial animal, the jaguar, can survive here year around because it is such a good swimmer. It may be ominous news that since designation of the reserve in 1990, jaguar numbers have increased and so have the problems. Attacks on cattle have mounted, and stockmen must bear the losses. In 1999, at least fourteen cats were known to have been killed with shotguns and fishermen's harpoons in retaliation. The situation illustrates how much we still have to learn about the New World's largest feline.

Above: *The Pixaim River in Brazil's Pantanal flows through good jaguar country, exquisite here at dawn during the dry season.*

Opposite page: *Jaguars are the third largest and the stockiest of all the world's big cats.*

Above: *In Central America, the common iguana is easy prey for jaguars.*

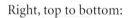
Right, top to bottom:
Caimans, especially small ones that are easily caught, are a diet staple in the Venezuelan llanos and elsewhere in jaguar range.

The Brazilian tapir is the largest prey commonly taken by jaguars.

Especially during dry seasons when herds of them travel to find water, capybaras are a staple of jaguars.

Many a coati makes an easy, although small, meal for a hunting jaguar.

Opposite page: *Despite its size, the spotted jaguar can stalk silently in almost any kind of cover.*

Above: *Jaguars are not often seen anywhere by humans, but the Cuiaba River vicinity in Brazil's Pantanal, may be the one best place.*

Opposite page: *Like so many wild felines, jaguars have long been trapped, hunted, and poached for their beautiful spotted pelts.*

Top: *The courtship and mating of jaguars is typically catlike—noisy and sometimes prolonged.*

Above: *A few jaguars may still cling to existence in the brittle, arid region along the United States–Mexico border. This is a young cub.*

Left: *An ocelot, a much smaller spotted cousin of the jaguar, catches an iguana.*

Above: *The ocelot,* Felis pardalus, *shares much of the jaguar's entire range and may be mistaken for a young jaguar.*

REFERENCES

Adamson, George. *A Lifetime With Lions*. Garden City, NY: Doubleday, 1968.

Ammann, Katherine, and Karl Ammann. *Cheetah*. New York: Arco Publishing, 1985.

Bauer, Erwin, and Peggy Bauer. *Big Game of North America*. Stillwater, Minn.: Voyageur Press, 1997.

Bauer, Erwin. *Treasury of Big Game Animals*. New York: Outdoor Life, 1972.

Bauer, Peggy, and Erwin Bauer. *Wild Kittens*. San Francisco: Chronicle Books, 1995.

Brakefield, Tom. *Big Cats: Kingdom of Might*. Stillwater, Minn.: Voyageur Press, 1996.

Busch, Robert H. *The Cougar Almanac*. New York: Lyons & Burford, 1996.

Caputo, Phillip. *Ghosts of Tsavo: Stalking the Mystery Lions of East Africa*. Washingtion, D. C. : National Geographic Adventure Books, 2002.

Corbett, Jim. *Man-Eaters of Kumaon*. London: Oxford University Press, 1944.

———. *The Man-Eating Leopard of Rudraprayag*. London: Oxford University Press, 1948.

Dalrymple, Byron. *North American Big Game Animals*. Harrisburg, Pa.: Stackpole Press, 1985.

Estes, Richard D. *Behavior Guide to African Mammals*. Berkeley: University of California Press, 1991.

Guggisberg, C. A. W. *Wild Cats of the World*. New York: Taplinger Publishing, 1975.

———. *Simba*. Philadelphia: Chilton Books, 1963.

Hillard, Darla. *Vanishing Tracks: Four Years Among the Snow Leopards of Nepal*. New York: Arbor House Publishing, 1989.

Hornocker, Maurice, editor. *Track of the Tiger: Legend and Lore of the Great Cat*. San Francisco: Sierra Club Books, 1997.

Hunter, J. A. *Hunter*. New York: Harper & Brothers, 1952.

Joubert, Dereck, and Beverly Joubert. *Hunting with the Moon: The Lions of Savuti*. Washington, D. C.: National Geographic Society, 1997.

Kat, Pieter W. *Prides: The Lions of Moremi*. Washington, D.C.: Smithsonian Institution Press, 2000.

Kitchener, Andrew. *The Natural History of the Wild Cats*. Ithaca, NY: Comstock Publishing, 1991.

Leopold, Starker. *Wildlife of Mexico*. Berkeley: University of California Press, 1959.

Lydekker, Richard. *The Game Animals of India, Burma, Malaya and Tibet*. London: Rowland Ward, 1907.

Matthiessen, Peter. *Tigers in the Snow*. New York: North Point Press, 2000.

National Wildlife Federation. *Kingdom of Cats*. Washington, D.C.: NWF Books, 1987.

Patterson, Lt-Col. J. H. *The Man-Eaters of Tsavo and Other East-African Adventures*. New York: The Macmillan Co., 1927.

Polking, Fritz. *Leoparden*. Steinfurt, Germany: Teklenborg Verlag, 1995.

Prater, S. H. *The Book of Indian Animals*. Bombay, India: The Bombay Natural History Society, 1965.

Rabinowitz, Alan. *Chasing the Dragon's Tail: The Struggle to Save Thailand's Wild Cats*. New York: Doubleday, 1991.

———. *Jaguar*. New York: Arbor House Publishing, 1991.

Savage, Candace. *Wild Cats*. Vancouver, B.C.: Douglas & McIntyre, 1993.

Schaller, George B. *Mountain Monarchs: Wild Sheep & Goats of the Himalaya*. Chicago: University of Chicago Press, 1977.

———. *Stones of Silence: Journeys in the Himalaya*. New York: Viking Press, 1979.

———. *The Deer and the Tiger*. Chicago: University of Chicago Press, 1967.

———. *The Serengeti Lion: A Study of Predator–Prey Relations*. Chicago: University of Chicago Press, 1977.

Seidenstricker, John. *Tigers*. Stillwater, Minn.: Voyageur Press, 1996.

Server, Lee. *Lions, King of Beasts*. New York: Todtri, 1993.

———. *Tigers*. New York: Todtri, 1991.

Singh, Arjan. *Tiger Haven*. New York: Harper & Row, 1973.

Sunquist, Fiona, and Melvin E. Sunquist. *Tiger Moon: Tracking the Great Cats in Nepal*. Chicago: University of Chicago Press, 1988.

Thapar, Valmik. *The Tiger's Destiny*. London: Kyle Cathie, 1992.

Thompson, Sharon. *Built for Speed: The Extraordinary, Enigmatic Cheetah*. Minneapolis: Lerner Publications, 1998.

Tinker, Ben. *Mexican Wilderness and Wildlife*. Austin: University of Texas Press, 1978.

Turnbull-Kemp, Peter. *The Leopard*. Cape Town, South Africa: Howard Timmins, 1967.

Ward, Geoffrey C. *Tiger Wallahs*. New York: Harper-Collins, 1993.

Wrogeman, Nan. *Cheetah Under the Sun*. Johannesburg, South Africa: McGraw-Hill, 1975.

Young, Stanley, and Edward Goldman. *The Puma*. New York: Dover Publications, 1946.

RESOURCES

African Wildlife Foundation
1400 16th Street NW
Suite 120
Washington, D. C. 20036 USA
www.awf.org

AfriCat
PO Box 1889
Otjiwarongo, Namibia, Africa
www.africat.org

Big Cats Research
www.bigcats.com

Born Free Foundation
3 Grove House
Foundry Lane
Horsham, West Sussex
RH13 5PL Great Britain
www.bornfree.org.uk

Cheetah Conservation Fund
PO Box 1380
Ojai, CA 93024 USA
www.cheetah.org

De Wildt Cheetah and Wildlife Centre
PO Box 16
De Wildt 0251 South Africa
www.dewildt.org.za

Defenders of Wildlife
1101 14th Street NW #1400
Washington, D. C. 20005 USA
www.defenders.org

Hornocker Wildlife Institute
2023 Stadium Drive, Ste. 1A
Bozeman, MT 59715
www.hwi.org

International Snow Leopard Trust
4649 Sunnyside Avenue N
Seattle, WA 98103 USA
www.snowleopard.org

International Society for Endangered
  Cats, Inc.
3070 Riverside Drive, Suite 160
Columbus, OH 43221 USA
www.isec.org

International Society for Endangered
  Cats–Canada
124 Lynnbrook Road SE
Calgary, AB T2C 1S6 Canada
www.wildcatconservation.org

Lion Research Center
100 Ecology Building
University of Minnesota
1987 Upper Buford Circle
St. Paul, MN 55108 USA
www.lionresearch.org

The Mountain Lion Foundation
PO Box 1896
Sacramento, CA 95812 USA
www.mountainlion.org

Mountain Lion Foundation of Texas
1715 Horton Preiss Road
Blanco, TX 78606 USA
www.mountainlions-texas.org

National Wildlife Federation
1110 Wildlife Center Drive
Reston, VA 20190-5362 USA
www.nwf.org

The Nature Conservancy
4245 North Fairfax Drive, Suite 100
Arlington, VA 22203-1606 USA
www.nature.org

Rhinoceros and Tiger Conservation
  Fund
U.S. Fish and Wildlife Service
http://international.fws.gov

Save the Tiger Fund
National Fish and Wildlife Foundation
1120 Connecticut Avenue NW,
  Suite 900
Washington, D. C. 20036 USA
www.nfwf.org/programs/stf.htm

Sierra Club
85 Second Street
San Francisco, CA 94105 USA
www.sierraclub.org

The Wilderness Society
1615 M Street NW
Washington, D. C. 20036 USA
www.wilderness.org

Wildlife Conservation Society
2300 Southern Boulevard
Bronx, NY 10460 USA
www.wcs.org

World Wildlife Fund/World Wide Fund
  for Nature
www.wwf.org

Index

The bleached skull of a marsh deer is all that remains of a jaguar kill in Brazil's Pantanal.

About the Photographers

From bases in Jackson Hole, Wyoming, Montana's Paradise Valley, and Washington State, Erwin and Peggy Bauer have been writing about and photographing the wild world for an aggregate of eighty years. They may be the world's most frequently published wildlife photographers. Cameras in hand, they have photographed from the Arctic Ocean to Antarctica, Yellowstone to Thailand, from Africa to numerous wild island archipelagos. They have published more than fifty books on natural history, including *Antlers, Whitetails, Mule Deer, Elk, Bears, Big Game of North America*, and *Yellowstone*, all published by Voyageur Press. Their work appears often in *National Wildlife, Smithsonian, Natural History, Sierra, Outdoor Life, Outdoor Photographer*, and many other publications. The Bauers were awarded the 2000 Lifetime Achievement Award by the North American Nature Photographers Association. They currently live on Washington's Olympic Peninsula, between the Olympic Mountains and the sea.